RED 5

AN INVESTIGATION INTO THE DEATH OF FLIGHT LIEUTENANT SEAN CUNNINGHAM

David Hill

Nemesis Books

ISBN 978-1-7061-4923-1 (paperback)

Originally published August 2019
(Revised and updated, July 2022)

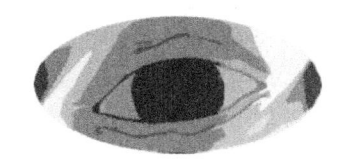

The right of David Hill to be identified as author of this work has been asserted by him under the Copyright, Designs and Patent Act 1988.

The views expressed are those of the author alone and should not be taken to represent those of Her Majesty's Government, the Ministry of Defence, HM Armed Forces or any Government agency, unless quoted. The events outlined are real. Other participants may have their own views.

Please feel free to reproduce or transmit this publication in any form, and by any means.

By the same author, in paperback and Kindle, and published by Nemesis Books:

Their Greatest Disgrace - The campaign to clear the Chinook ZD576 pilots

ISBN 979-8-8429-8674-3 (2016)

Breaking the Military Covenant - Who speaks for the dead?

ISBN 978-1-9810-3842-8 (2018)

The Inconvenient Truth - Chinook ZD576: Cause & Culpability

ISBN 979-8-5184-5820-8 (2021)

A Noble Anger - The manslaughter of Corporal Jonathan Bayliss

ISBN: 979-8-8342-7923-5 (2022)

As the issues related in these books are ongoing, and they will be regularly updated. If you have purchased a previous edition, please contact the author for a free Kindle or pdf version. No proof of purchase required.

nemesisbooks@aol.com

https://sites.google.com/site/militaryairworthiness/home

The author is a retired aircraft engineer, spending much of his career as an avionics and aircraft project/programme manager in the Ministry of Defence. After retiring in 2004, he assisted families and Coroners in the Nimrod XV230 and Hercules XV179 cases. Also, Mr Charles Haddon-Cave QC's Nimrod Review (2008-9) and Lord Alexander Philip's Mull of Kintyre Review (2010-11).

In 2016 he published 'Their Greatest Disgrace', a personal account of the campaign to clear the Chinook ZD576 pilots (Mull of Kintyre 1994).

This was followed in 2018 by 'Breaking the Military Covenant', taking the 2003 Sea King ASaC Mk7 mid-air collision as its main test case, and linking the failures to over 110 other avoidable deaths sharing the same root causes.

In 2021 he published the prequel to Their Greatest Disgrace, 'The Inconvenient Truth - Chinook ZD576: Cause and Culpability'.

This book is dedicated to:
My daughter Jo and brother Chris
Diane Cottrell

'Eye' illustration by Kaelan Perrett

All proceeds to St Richard's Hospice, Worcester
https://www.strichards.org.uk/raise-funds/

Acknowledgements

Richard Anderson, engineer.

Allan Brocklebank, engineer.

Squadron Leader Robert Burke, test pilot.

Tony Cable, retired Senior Engineering Inspector, Air Accidents Investigation Branch.

Captain Robin Cane, pilot.

Commander Steve George RN, engineer.

Flight Lieutenant James Jones, engineer.

Sean Maffett, retired RAF navigator, now an aviation journalist.

Graham Miller, engineer.

Geoff Millward, engineer.

Dr Susan Phoenix

Dr Michael Powers QC

Christopher Seary, for kind permission to use the cover image.

Brian Wadham, engineer.

Table of contents

Introduction 1
Timeline 5

PART 1 - PROTECT AND SURVIVE 7
1: The Red Arrows 8
2: Martin-Baker Aircraft Company Ltd 11
3: The Martin-Baker Mk10 Ejection Seat 14

PART 2 - MAKE IT SAFE, KEEP IT SAFE 20
4: Hazards, Risks and Certainties 21
5: Safety 26
6: Accident causation & defences in depth 34

**PART 3 - THE ACCIDENT, MOD INVESTIGATION, AND
TECHNICAL MATTERS** 38
7: The Accident 39
8: The Problem 44
9: Applying the Routine Technical Instruction 55
10: Special information Leaflets (SIL) 704 & 704A 65
11: The Service Inquiry 68
12: The alleged lack of technical information 80
13: Airworthiness Review Teams, the Gas Shackle, and the 2002
QinetiQ report 100

PART 4 - THE PURSUIT OF MARTIN-BAKER 108
14: The Coroner's Inquest 110
15: The Health and Safety Executive 113
16: Lincolnshire Police & the Crown Prosecution Service 118
17: Court Appearances (2017) 122

18: Trial Day, 22 January 2018 125
19: Between trial & sentencing 128
20: Sentencing hearings, 12/13/23 February 2018 134
21: Sentencing remarks of the Honourable Mrs Justice Carr, DBE, 25 February 2018 145

PART 5 - FULL CIRCLE 156
22: Video evidence 157
23: Follow-up 160
24: A pattern of behaviour 169
25: Prevention is easier than the cure 174

ADDENDUM - NEW DAWN FADES 186
26: Conduct without conscience 187
27: Links to the Shoreham Air Display accident 190
28: Common factors between the Hillsborough tragedy and military accidents 198
29: The death of Corporal Jonathan Bayliss - Hawk XX224, RAF Valley, 20 March 2018 202

Glossary of terms and abbreviations 207

Figures

1. Bernard Lynch, BEM (1953).
2. Martin-Baker Mk10 ejection seat.
3. General view of Scissor and Drogue Shackle arrangement, with Scissor jaws retaining the Drogue Shackle.
4. Shackle Assembly in vertical position, with Restraining Plunger preventing Scissor from opening.
5. XX177 Drogue Shackle.
6. The Reason 'Swiss Cheese' Model of accident causation.
7. XX177 Drogue Shackle.
8. XX177 and sample Drogue Shackles, illustrating degree of over-tightening.
9. XX177 Scissor Shackle.
10. Mk10B seat, illustrating normal gap between shackles.
11. Mk10A ejection seat with Gas Shackle.
12. Still from 1959 RAF training film, showing shackle disengagement check.
13. Top Latch Plunger, showing correct and incorrect assembly.
14. HSE Principal Investigator demonstrating shackle disengagement.

Introduction

On 8 November 2011, Flight Lieutenant Sean Cunningham, 'Red 5', was due to fly Hawk T.1 XX177 from RAF Scampton in Lincolnshire as part of a Red Arrows training sortie, followed by a transit across England and Wales to RAF Valley on Anglesey.

At 1106, while on the ground with the engine running and canopy closed, his ejection seat fired. The sequence operated as expected up to and including deployment of the two drogue parachutes. However, the main parachute did not deploy and Flight Lieutenant Cunningham remained strapped in until impact with the ground.[1]

Red Arrows personnel, RAF Scampton emergency medical staff and civilian paramedics attended to him at the scene. He was transferred by air ambulance to Lincoln General Hospital, where he was pronounced dead. He had died at or just after impact.

As this was a 'zero-zero' seat (one designed to operate in a stationary ground level ejection), he would likely have survived had everything else worked normally. However, a series of Ministry of Defence (MoD) failures and violations culminated in a maintenance error by the RAF. A nut was over-tightened, preventing release of the main parachute.

*

In July 2013, MoD admitted liability. The Health and Safety Executive then prosecuted Martin-Baker Aircraft Company Ltd. It was alleged that, by virtue of queries raised by McDonnell Douglas in 1990 and British Aerospace in 1991, a warning should have been provided between October 1990 and July 1992 about the risk associated with over-tightening the nut.[2] It was further alleged that the information had <u>never</u> been provided, from the introduction of the design in 1952 to the date of the accident. Both allegations were known to be untrue.

Verbal, written and video evidence proved conclusively that Martin-Baker <u>did</u> provide the information necessary to ensure correct operation, initially in the 1950s. But MoD later instructed its maintainers not to use it. An important consideration is that MoD, as we know it, was

---

1 https://www.gov.uk/government/publications/service-inquiry-into-the-accident-involving-hawk-tmk1-xx177
2 McDonnell Douglas merged with Boeing in August 1997.

only created in 1964. From 1946 to 1964 a number of departments did the work of the modern MoD, including the Admiralty, War Office, Air Ministry, Ministry of Aviation, and an earlier form of MoD. They merged in 1964. The defence functions of the Ministry of Aviation Supply then transferred in 1971. That is, between 1952 and 1971 information was provided to departments that no longer exist. The responsibility to retain and maintain it lay with MoD, not the company.

Despite this exculpatory evidence (that which exonerates), the company pleaded guilty, and on 23 February 2018 were fined £1.1M. Those responsible for Sean Cunningham's death have yet to be called to account.

*

The main public record is the Service Inquiry report. It is a notable work, but the Panel was constrained by its terms of reference, and misled by other parts of MoD. I must emphasise that I differentiate between the Panel, and the Military Aviation Authority who issued the report. The former's role was to address this one accident. The latter is aware of the linkages with others, and is, as I shall explain, culpable.

Here, Part 1 is available in a redacted form. Short, heavily redacted extracts from Part 2 were belatedly released under Freedom of Information. I leave it to you whether to download the report. I recommend you do, but read it alongside another. You will see, despite its limitations, it is more candid than most.

*

After discussing what the report says, I move on to what it omits. This forms the bulk of my evidence. My starting point is:

- Martin-Baker certified the seat safe, if serviced correctly.
- Hawker Siddeley (later part of British Aerospace) certified it safe for use in Hawk T.1 aircraft, subject to the same caveat.
- Predicated on its ability to conduct this servicing in accordance with Martin-Baker instructions, MoD issued the Release to Service (the formal name for the Master Airworthiness Reference) permitting Service regulated flying. MoD then introduced various policies *prohibiting* correct servicing. Henceforth, a prerequisite to airworthiness certification was missing.

I follow a broadly chronological sequence: background, the accident, investigations, legal process and analysis. To help place matters in context, I offer a flavour of events and MoD policies at each key stage. A little understanding is required of MoD's Safety Management System - I limit myself to the most basic part of the airworthiness chain: Statement of Operating Intent and Usage » Build Standard » Safety Case » Release to Service.

Some technical and procedural detail is necessary, mainly because no investigation sought to understand root causes. (A factor is considered a root cause if its removal from the problem-fault sequence would have prevented the undesirable outcome). Investigations were confined to the final acts - the inadvertent ejection and failure of the main parachute to deploy - leaving many questions hanging, which I seek to answer.

The section dealing with Lincolnshire Police and the Crown Prosecution Service is short. Both were found wanting, disregarding serious offences.

The Health and Safety Executive did not publish a report, but its position is known through court records and correspondence. Its actions served to place aircrew at greater risk, because the facts were suppressed making recurrence more likely.

These investigations were chronologically and causally wrong.

*

My analysis of court proceedings draws on contemporaneous notes. I offer proof that the Coroner and Judge were misled by omission and commission, distorting their views. Hence, the Cunningham family, servicemen, media and public were also misled. I then provide a detailed assessment of the Judge's sentencing remarks, which were based on flawed and false evidence. Nevertheless, they reveal she knew who was truly culpable, and I discuss the dilemma this presented her; one she acknowledged caused great difficulty.

I address a major question - why did Martin-Baker plead guilty, knowing they were innocent? The answer leaves a bad taste, affording MoD a degree of protection it normally only receives from legal authorities and Parliament.

What I will not do is criticise aircrew, maintainers or the Red Arrows. While there were certainly errors, MoD used them to divert attention from the difficulty it faced if the facts were revealed. Primarily, that the root causes had all been notified, and ignored for well over two decades. It knew this was a recurrence. Only the detail of the final acts differed. In all other respects, Sean Cunningham was a victim of the same violations that had already killed many of his colleagues.

MoD's Safety Management System had collapsed, long before 2011. RAF Directors of Flight Safety had, throughout the 1990s, warned senior staff of systemic failures and standing risks. Of particular relevance, it was known deficiencies existed with engineering training and maintenance publications. The RAF's reaction was to make further cuts and continue with its policy of *savings at the expense of safety*. Safety margins across military aviation were consciously and relentlessly eroded to an unacceptable and dangerous degree.

This accident, like so many others, would have been avoided by implementing mandated regulations. False declarations were made that they *were* implemented. These offences were advised to Ministers. When they sought MoD briefings, they were misled and lied to, causing Parliament to be misled.

The inevitable question arises. Who can one trust?

Timeline

The case is characterised by a series of incorrect claims as to timing of events. Here is a summary of significant dates. For now, I ask you to accept these are verifiable facts. Later, I will explain each.

- **1952** Scissor/Drogue Shackle parachute release mechanism introduced with Mk2 ejection seat. The same design arrangement remains in the Hawk Mk10B seat. *The Health and Safety Executive denies this.*
- **1953** RN maintainers trained by Martin-Baker to upgrade existing seats to Mk2, incorporating the new release mechanism.
- **1955** RAF training includes appropriate instructions for disassembly and reassembly of ejection seats, requiring shackle disengagement checks to be carried out in a servicing bay before declaring the seat serviceable and installing it. *The Health and Safety Executive denies this.* Checking disengagement is not possible in the aircraft.
- **1959** The above is reflected in RAF training films. *The Health and Safety Executive denies this.*
- **1984** Gas System Drogue Bridle release system designed. MoD aware it eliminates the risk of shackle pinching by removing the Scissor Shackle. *The Health and Safety Executive denies this.* Thereafter, this is the standard, recommended design. MoD retro-fitted Tornado, but not Hawk.
- **August 1991** MoD signs off a Martin-Baker report, confirming it is content with how the company managed customer queries regarding the parachute release mechanism. Martin-Baker later prosecuted for not informing MoD of issues the report addresses. *The Health and Safety Executive denies that MoD signing and accepting this report is relevant.*
- **15 November 2011** One week after the accident, Martin-Baker issue a Special Information Leaflet reiterating how to ensure the main parachute deploys correctly. MoD continues to ignore specific warnings contained therein, and issues an opposing instruction.
- **13 April 2013** Crown Prosecution Service issues a statement saying it will bring no charges against MoD or Martin-Baker.

- **July 2013** MoD admits liability and later that year agrees a settlement with the Cunningham family, subject to a non-disclosure agreement.
- **January 2014** Coroner's Inquest. False allegations repeated.
- **26 September 2016** Health and Safety Executive announces its intention to prosecute Martin-Baker.
- **28 September 2016** Health and Safety Executive informed of exculpatory evidence, but does not interview witnesses or review that evidence.
- **25 January 2018** Health and Safety Executive informed of further exculpatory evidence, in the form of independent reports. It denies their relevance.
- **6 February 2018** Coroner Stuart Fisher sends the same evidence to the Judge, Mrs Justice Carr.
- **23 February 2018** Martin-Baker sentenced and fined.
- **22 May 2018** Formal complaint made to Lincolnshire Police that (a) evidence was not disclosed by MoD, and (b) when it was uncovered, the Health and Safety Executive did not investigate, thus misleading the court. Lincolnshire Chief Constable refuses to investigate, ruling that MoD and the Health and Safety Executive may judge their own case.

*

1952 may seem a long time ago, but my timeframe was set by the Health and Safety Executive's claim that Martin-Baker <u>never</u> provided the necessary information. It offered no supporting evidence, and after sentencing admitted it held none.

On 20 September 2018, my manuscript was forwarded to the Cunningham family via their solicitor.[3] The content has also been sent to MoD, the Coroner, Lincolnshire Police, the Health and Safety Executive and the Judge. My approach is 'silence gives consent'.

---

3 Confirmatory e-mail from Keith Barrett, solicitor, 20 September 2018 22:36.

PART 1 - PROTECT AND SURVIVE

'Any appliance of a mechanical nature will be more reliable and remain efficient longer if it is maintained and serviced correctly at frequent intervals by personnel familiar with that device. An ejection seat, being a particularly important piece of mechanical equipment apparatus for saving lives, is no exception to this rule and, as such, it was thought advisable to establish a training school at which servicing personnel and aircrew could familiarise themselves with the workings and maintenance of the Martin-Baker seat'.

Martin-Baker 'School for Safety' (established 1955)

1: The Red Arrows

The Red Arrows, officially the Royal Air Force Aerobatic Team, are the world-renowned display team of the RAF, based at RAF Scampton, near Lincoln.

Since 1920 a number of RAF teams have flown air displays, using aircraft ranging from biplanes to fast jets. They were amalgamated in 1964, the Red Arrows beginning life at RAF Fairford in Gloucestershire, then a satellite of the Central Flying School. They moved to RAF Kemble, now Cotswold Airport, in 1966. When Scampton became the Central Flying School headquarters in 1983, the team moved there. As a savings measure, Scampton closed in 1995, the team moving 20 miles to RAF Cranwell. However, as they still used the air space above Scampton, the emergency facilities and runways had to be maintained. On 21 December 2000, they returned. In July 2018, it was announced Scampton was to close; the team due to relocate to RAF Waddington by end-2022. The future is uncertain.

*

Initially, the team was equipped with seven Gnat trainers inherited from the RAF Yellowjacks display team. In 1968 this was increased to nine, and in 1979 it took delivery of Hawks.

The Hawk T.1 is a flying training and weapons training aircraft. Equipment specific to the weapons training role is fitted only when aircraft are allotted to that role. The Hawk T.1A is equipped to an operational standard capable of undertaking a war role; the equipment for this role is also suitable for the weapons training role. The Red Arrows fly both variants.

The two seats are positioned in tandem, with the handling pilot in front and any passenger/instructor behind. (In certain training scenarios the instructor will be in the front seat). The rear seat occupant is not a 'co-pilot'. While the aircraft can be flown from the rear, the cockpits are not identical, and nor are the seats. For example, the rear occupant cannot select radio frequencies, operate the TACAN (Tactical Air Navigation System) or Instrument Landing System. Additionally, the IFF/SSR (Identification Friend or Foe/Secondary Surveillance Radar) is controlled from the front seat, providing identification and coded flight level information to interrogating stations or aircraft.

The basic armaments installation provides for the carriage and release of rockets and practice bombs on two wing pylons, and for a 30mm ADEN cannon carried in a pod on the fuselage centreline. Additionally, the T.1A is equipped for AIM-9 Sidewinder air-to-air missiles to the wing pylons.

While these details are irrelevant in this case, the pods affect the handling of the aircraft; which was a factor in another accident discussed, the death of Corporal Jonathan Bayliss in March 2018.

In more general terms, much of the avionic fit in the Hawk is regarded as, at best, obsolescent by the rest of MoD. For example, its Standby UHF emergency radio was declared obsolete by the Fleet Air Arm in 1990. So old is the technology, it is extremely difficult to support, with spares availability poor. More importantly, the malaise of under-investment militates against adequate training which, as we will learn, was a major factor in the accident and death

*

A subtlety here is that the Red Arrows are now the only squadron with all-military aircrew and engineers. That is, what would be considered a 'traditional' squadron, whereas others are now supported by outsourced maintenance contracts (perhaps MoD's most self-destructive decision), and have embedded contracted pilots.

This brings its own problems, the worst (to me, as an engineer), being they do not have their own Engineering Wing. So, their Senior Engineer (SEngO) is also their *de facto* Officer Commanding Engineering and Supply. Also, being remote from other Hawk squadrons means they must, for example, travel from Lincolnshire to Anglesey (RAF Valley) to carry out simulator training. In short, the squadron is poorly served, which may seem odd to an outsider reading this. In 1992 the RAF Director of Flight Safety characterised the Chinook as the *'Cinderella'* of the RAF fleets. The Hawk is its twin sister.

*

All pilots are volunteers. They must have completed one or more operational tours on a fast jet, have accumulated at least 1,500 flying hours, and been assessed as above average in their operational role. Applications are always oversubscribed. They stay with the team for a 3-year tour. Three are changed every year, such that normally three first-year, three second-year, and three in their final year are on the team. The team leader, Red 1, also spends three years there. He is always a pilot who has previously completed a tour.

Red 10, the team supervisor, is a fully qualified Hawk pilot and flies a tenth aircraft when the team is away from base; meaning there is a reserve aircraft at the display site. His duties include co-ordination of all practices and displays, and acting as the team's ground safety officer.

The squadron is the window through which everyone else views the RAF. It is important that the standards applicable to the wider RAF are seen to apply to the Red Arrows. They must stand out because they are the best, not because they do things differently.

2: Martin-Baker Aircraft Company Ltd [4]

'It is not enough to reflect with pride on the number of lives saved, but more worthwhile to consider those ejections which were not successful, to discover the reason why and, if possible, to remedy this in future seats'.

Editorial, Martin-Baker Review #3, December 1960

In August 1929 James Martin came to Denham, Buckinghamshire, to build aeroplanes that were to be different, simpler, cheaper and easy to produce and maintain. He had two employees. In 1934, he and his friend Captain Valentine Baker formed Martin-Baker. A series of innovative designs followed. Not only aeroplanes. A balloon cable cutter. A new jettison hood for Spitfires. A belt-feed mechanism for aircraft-mounted machine guns. A wind indicator to assist pilots. An automatic centrifugal oiling unit for propellers.

As speeds increased, so too did the difficulty in simply bailing out. On 12 September 1942 Captain Baker died during a test flight of a prototype MB3 aircraft, the engine seizing and the aircraft striking a tree stump during the emergency landing. Greatly affected, Mr Martin turned his mind to pilot safety. In 1944 he was invited by the Ministry of Aircraft Production to investigate the practicality of providing fighter aircraft with a means of assisted escape for the pilot. After considering various schemes, it became apparent the most attractive means would be by forced ejection using an explosive charge. The Martin-Baker ejection seat was born. The first emergency ejection using a Martin-Baker seat was on 30 May 1949, by Armstrong-Whitworth test pilot Flight Lieutenant John Lancaster DFC, from a prototype AW52 Flying Wing.

The process was risky and experimentation sometimes dangerous. The ability of humans to tolerate 'g' forces had to be assessed. A seat had to be designed to hold the occupant sufficiently stable to ensure adjacent spinal vertebrae were square to each other, while controlling peak acceleration. In August 1945, one employee, experimental engineer Bernard (Benny) Lynch, undertook the first live 'ride' up the test rig. A further 179 rig ejections were carried out. One of the first serving officers to do so was Air Commodore Eric Lumley, Fighter Command's Principal Medical Officer.

---

4 Includes extracts from 'The Story of an Enterprise' (1955), by James Martin.

Successful air testing followed, in ever faster aircraft. A dummy was ejected from a Meteor at 415mph in June 1946. Mr Lynch made the first live ejection on 24 July 1946, from 8,000 feet at 320mph. At the time, perceived wisdom was that ejection at over 300mph was unsurvivable. He undertook the test knowing he had to remain conscious, unfasten himself from his restraining harness, pause for a few seconds to avoid any risk of entanglement, and pull the ripcord. This was bravery of the highest order, and one of the great landmarks in aviation history. On 19 August 1947, he ejected from 12,000 feet at 420mph. Mr Lynch continued his work with the company and later successfully ejected at 30,000 feet. This remains one of the highest ever.

In 1947 it was decided to standardise the Martin-Baker design in all new military jet aircraft. The Mk1 seat was born, fitted to Canberra, Meteor, Sea Hawk, Venom and others. It soon became apparent an automatic system of deployment was required to cater for aircrew incapacitation. A Barostatic Time Release Unit and Scissor/Drogue Shackle Assembly was developed in 1949, entering Service in 1952. That primary design remains in Red Arrows Hawks - an indisputable fact that is the central issue of this case, but which the Health and Safety Executive denies.

James Martin was knighted in 1965, and passed away in 1981. His sons and grandsons continue his pioneering work. At time of writing, over 7,600 airmen owe him their lives. Of these, over 800 are RAF aircrew.

Figure 1 - Bernard Lynch BEM (1918-86) on 17 March 1953, prior to undertaking an ejection at 30,000 feet from a Gloster Meteor Mk3. The seat used was a modified Mk1, with a prototype automatic parachute deployment mechanism. *(Martin-Baker)*

3: The Martin-Baker Mk10 Ejection Seat

An ejection seat is a device whose design must constantly evolve with the desire for faster, higher and more manoeuvrable aircraft. There is much to consider. In addition to the ejection itself, escape path clearance, sequencing, stabilisation, life support, descent, rescue, and more. One cannot demonstrate design maturity through regular in-service use as quickly as the rest of the aircraft. In fact, one hopes never to. Normally, this is an iterative process, with numerous repeat cycles to analyse. With an ejection seat these are limited and irregular, with infinitely variable scenarios and outcomes. And the results of one iteration are not necessarily the input to the next. Effectively, the design is perpetually immature. Physical space in the aircraft, and the extremely short decision-making and execution process, militate against back-up systems.

It is vital not to underestimate the psychological effect of knowing one is sitting in the world's best and most reliable escape apparatus. The seat is the last resort, employing hazardous and violent methods when the aircraft is normally already lost and the sole purpose is to try to save the occupant. It is the final, magnificent, defiant defence against certain death. When a seat fires, it means all other defences have been breached. In this case, and on its own admission, by MoD - not Martin-Baker. The death of Sean Cunningham differs only in being an uncommanded ejection. The principles remain the same.

Set against this background, Martin-Baker's contribution to safety is unparalleled in the history of aviation - testament to the enduring quality of its designs. That legacy will stand for all time.

General description

The Mk10 seat is designed in four main units: ejection gun, main beam structure, seat pan, and parachute assembly. (MoD documentation incorrectly classifies the parachute assembly as part of the main beam structure - a crucial error).[5] This simplifies and speeds maintenance or cockpit access. For example, the seat pan can be removed to provide full access to the equipment in the cockpit, or aid removal of dropped objects, without disarming or removing the seat main structure or

---

5 AP 101B-4401-15, Part 1, Chapter 9.6. Hawk Aircrew Manual.

aircraft canopy.

However, this simplicity of design does not remove the obligation to observe basic safety procedures, particularly the need to avoid *in situ* seat maintenance if serviceability cannot be verified. I labour this point. Failure to heed it killed Sean Cunningham.

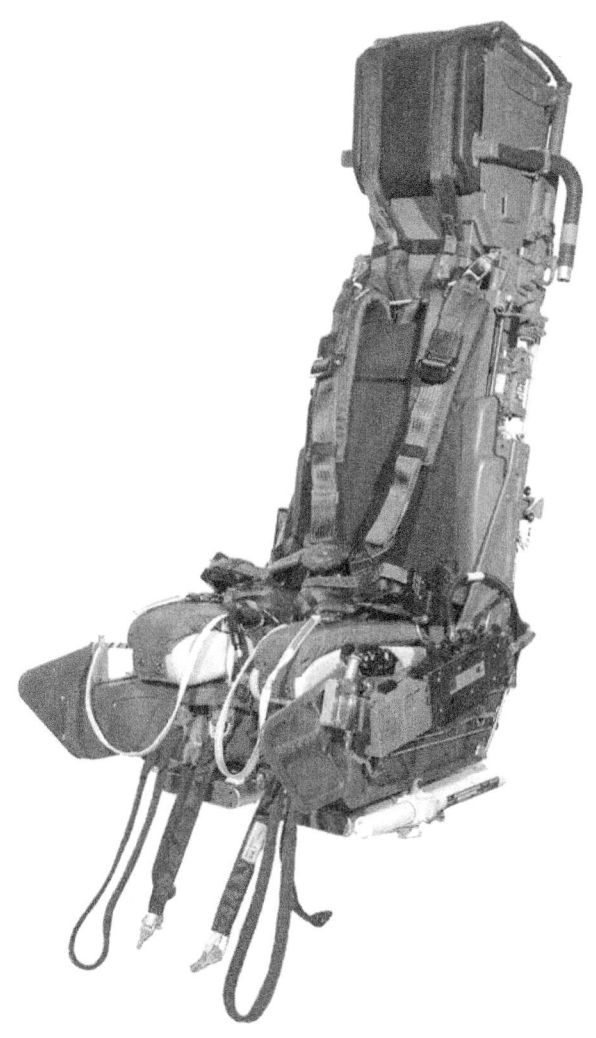

Figure 2 - Martin-Baker Mk10A Ejection Seat *(Martin-Baker)*

Basic operation

The following is applicable when only the pilot is in the aircraft. Matters are slightly different if there are two occupants, but that is not pertinent here.

The pilot is held in by straps. If it is necessary to abandon the aircraft he pulls a handle situated between his legs (the Seat Pan Firing Handle), initiating the ejection process. A complex and fully automatic train of events takes place in a matter of seconds.

Briefly... An explosive cord is detonated, shattering the Perspex canopy above and around the pilot. (If the detonating cord fails, the seat will penetrate the canopy without affecting seat performance). The ejection gun fires, propelling the seat upwards along extending rails. A series of rockets are activated, rapidly taking the seat, with the occupant still strapped in, up and out of the aircraft. Two drogue parachutes are released, termed the Duplex Drogue Assembly. The smaller, the Controller Drogue, stabilises the seat in a horizontal attitude, so the subsequent deceleration of the seat by the Main Drogue is more tolerable. At a predetermined altitude the main parachute deploys. The pilot is separated from the seat and returns to the ground, suspended under the main parachute. The seat descends separately.

The parachute release mechanism

The two drogue parachutes and main parachute are housed in a rigid container (the Head Box) on the top of the seat. They are connected to a Drogue Shackle, which is held by a restraining Scissor Shackle, which is permanently bolted to the ejection seat. (Figure 3).

When the firing handle is pulled, a drogue weight is fired into the air. It is attached to the smaller of the drogues; in turn attached to the larger one; and it to the main parachute.

The weight pulls out the smaller drogue, which inflates, pulling out the larger one. The drag of the drogues causes the Scissor Shackle, with the Drogue Shackle attached, to move like a hinge from a horizontal to vertical position. (Figure 4).

Initially, the Scissor Shackle jaws hold the Drogue Shackle in place around the Drogue Shackle bolt, preventing disengagement and release of the main parachute.

As the seat descends through 10,000 feet, the Barostatic Time Release

Unit operates. A rod is withdrawn from a restraining plunger, allowing the Scissor Shackle to open. (Figure 4).

This releases the Drogue Shackle, uncoupling the drogue parachutes from the seat. They continue to pull on the now free Drogue Shackle, pulling out the main parachute. The seat harness is released and the occupant is lifted out of the seat.

If the ejection occurs over high ground (>10,000 feet), the occupant can manually initiate an emergency release. If the ejection takes place below 10,000 feet, including while stationary at ground level, the sequence of events is continuous as the Barostatic Time Release Unit operates immediately.

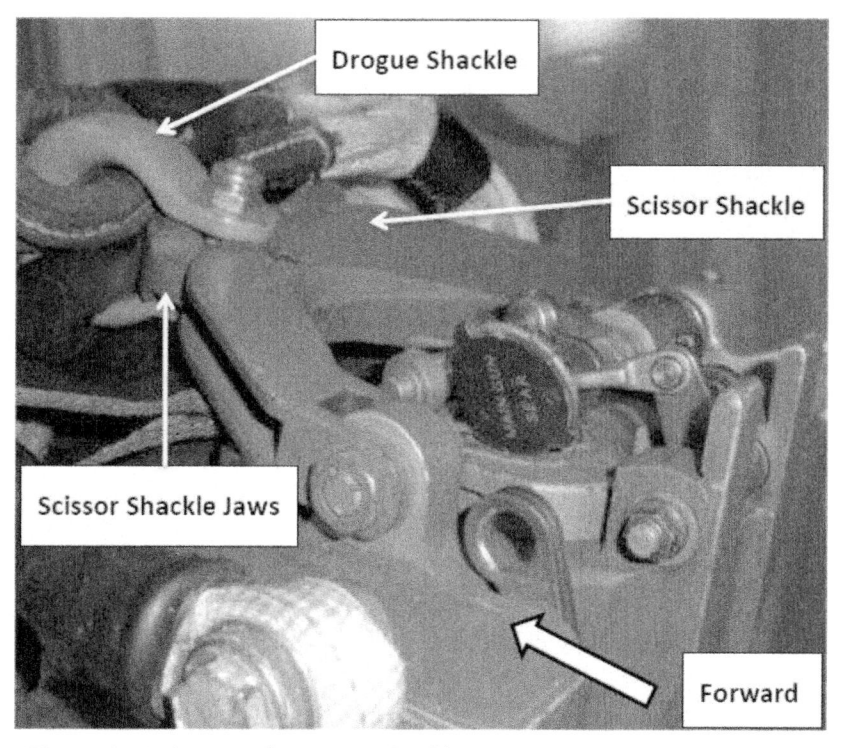

Figure 3 - Scissor and Drogue Shackle arrangement, with the Scissor jaws closed around the Drogue Shackle bolt. *(Public domain)*

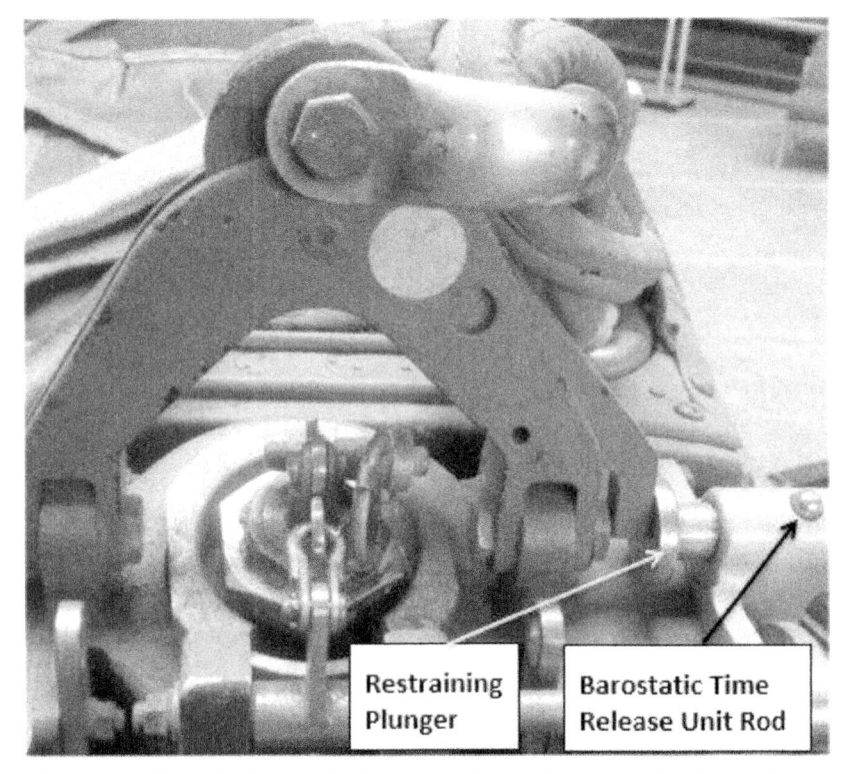

Figure 4 - Shackle Assembly in vertical position, with the Restraining Plunger preventing the Scissor from opening. *(Author)*

If the two shackles pinch together, they cannot disengage (or disengagement is delayed, depending on how severe the pinching is) and the main parachute will not deploy.

Thus, clearly, the final functional check on the release mechanism is to verify disengagement when the Scissor is opened. <u>Once reassembled, the shackles must not be disturbed without re-checking this</u>. In this case they were - the fatal error. I discuss this later in more detail.

Drogue Shackle and disengagement

By design, the Inner Lug Gap is considerably wider than the thickness of the Scissor Shackle jaws. (Figure 5). When assembled correctly, the Drogue Shackle is free to swivel about the bolt, and disengage when the jaws open. Aircrew can check the former by 'rattling' the Shackles. <u>But</u>

not the latter. It is for groundcrew to ensure the gap remains correct when assembled, by (a) not over-tightening the nut, and (b) checking shackle disengagement. The way one ensures the gap is correct is to follow Martin-Baker's instructions, which are to tighten the nut until it just touches the lug. That is, do not torque load the nut, and do not compress the lugs.

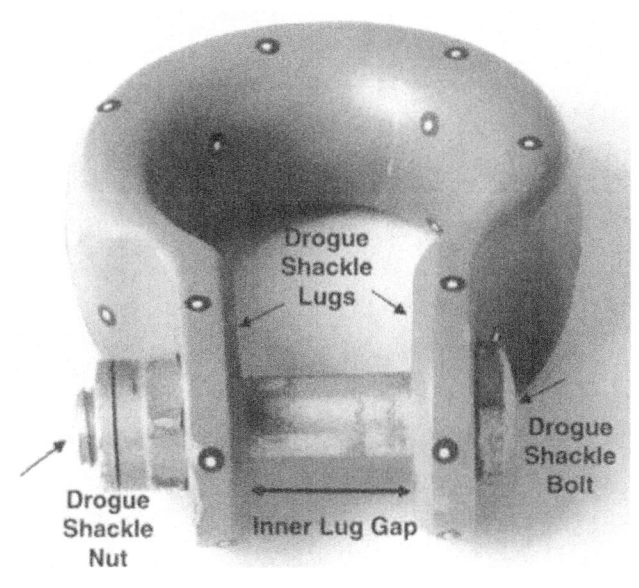

Figure 5 - XX177 Drogue Shackle. Note: the lugs are not parallel, having been compressed by an over-tight nut. *(MoD)*

The Service Inquiry reported that during the application of Routine Technical Instruction/Hawk/059 on 24 October 2011, RAF groundcrew over-tightened the Drogue Shackle nut, causing the Inner Lug Gap to be too small and pinch on the Scissor Shackle. The shackles could not disengage, and the main parachute did not deploy. The Panel replicated this scenario, determining the pull weight necessary to disengage the shackles in XX177. It concluded the aircraft would have had to be travelling in excess of 50 knots for the drogues to provide sufficient drag. However, it omitted that Sean's seat had not been checked for shackle disengagement, so had not passed a valid serviceability test. It was therefore not permitted in the aircraft. Its documentation was signed to say it *was* serviceable.

PART 2 - MAKE IT SAFE, KEEP IT SAFE

'Anomalies in testing should bring your organisation to a standstill. They are a violation of requirements. They are a clue something worse may happen'.

Dr Sheila M. E. Widnall, US Secretary of the Air Force and investigator of the Space Shuttle Columbia disaster.

4: Hazards, Risks and Certainties

A hazard is a potential source of harm. There are many types of impact, but here we are concerned with those that cause Risk to Life.

Risk Analysis is the identification and evaluation of the probability of occurrence and impact. Risk Assessment determines the significance and value of the identified hazards, and estimated risks to those affected. Risk Management is the monitoring and control of risk through development and implementation of mitigation plans. Continual assessment of mitigation action is necessary, along with contingency planning to cater for its failure, and the resources to manage this. Vital to this case, when known in advance a risk <u>will</u> occur, or pre-exists, it is a Certainty and <u>must</u> be dealt with before proceeding. Failure, in fact refusal, to do so has killed many, and was for example the primary root cause the Nimrod MRA4 programme cancellation.

A 'just' culture is necessary, where due regard is given to honesty but where individuals are not free from censure if they are culpably negligent. In recent years parts of MoD have tried this, only to be undermined by high-profile cases of leadership (and here, the Health and Safety Executive) knowingly placing blame on the wrong persons.

Sacrifice

Defence Standard 00-56 ('Safety Management Requirements for Defence Systems') states:[6]

'A system is safe when risk has been demonstrated to have been reduced to a level that is As Low As Reasonably Practicable [ALARP] and tolerable, and relevant prescriptive safety requirements have been met, for a system in a given application in a given operating environment. A risk is ALARP when it has been demonstrated the cost of any further risk reduction, where the cost includes the loss of defence capability as well as financial or other resource costs, is grossly disproportionate to the benefit obtained from that risk reduction'.

ALARP is about the practicality of risk reduction. Tolerability is about willingness to bear the risk. The link between the two being the sacrifice required to be made. In the military this involves placing a financial value on human lives, injuries, or environmental damage, so is an

---

[6] 'Requirements' are mandatory. It is important therefore that they are formalised.

emotive subject. An oft-overlooked subtlety is that while cost is a legitimate consideration, affordability is not.

While day-to-day management is delegated, at regular intervals the Operating Duty Holder (usually 2 Star level, an Air Vice Marshal or equivalent) must make a formal declaration that he has scrutinised residual risks and they are tolerable and ALARP. In theory this is legally binding, but has never been tested in court.

What is reasonable and tolerable? MoD issues guidelines, but they are just that. The enduring problem is the blurring of what is and is not practicable, and what is acceptable, caused by arbitrary and draconian cuts to safety related funding. 'Practicable' gets overtaken with *waste of money* and *don't do it*. 'Acceptable' with ambivalence, those issuing the instructions unprepared to place their reasons on record. The enduring question is - *Where does one stop?*

One could also ask - *Where do I begin?* An intolerable risk can be ALARP, since nothing more can be done. Service personnel face many scenarios where risk assessment is utterly futile. Often, an impossible dilemma is presented. To take this to an extreme, there are occasions when sending them to their certain death is not only tolerable, but demanded and accepted by society. That is the burden of command, and beyond the scope of this book. I'm not avoiding the issue. I've been in the position of 'managing' such a risk, where the Army dictated that one element of the design for a new piece of equipment, when operated remotely and correctly, would kill the user. Nothing in MoD regulations or guidelines helps in the slightest. But one thing is certain. You don't just sign. You go to see those involved and look them in the eye while explaining.

*

An intention to reduce a risk to ALARP, even if the required action is in hand, does not mean the risk *has* been reduced to ALARP. This is only achieved when the necessary control measures are in place. In recent years MoD has adopted the concept of 'ALARP (Temporal)', using this to justify carrying the highest level risks while mitigation is being considered, but not in hand. This may be acceptable given an operational imperative, but is routinely used to explain away decades-old risks which were understood from the outset and ignored. Mid-air collision prevention and fire-retarding technology are good examples; with high profile cases and major loss of life involving Tornado and

Hercules C-130 respectively.[7]

If safety management is not funded adequately, the extent of a Duty Holder's responsibility becomes unclear (as it is not they who request or refuse the funding), leaving everything at the door of the Secretary of State. In practice, he is never called to account, meaning some risks can never be controlled. Duty Holders then adopt different measures, ranging from grounding the aircraft to doing nothing. This case is an example of the latter. Staff who manage this are constrained by policies beyond their control. The question is doubly difficult because there is no independent oversight. Repeatedly, this has proved fatal.

Applicability to this case

If an ejection seat must be used, all defences against occurrence have failed. The seat is only required to work once, and must do so successfully. Correct maintenance is critical. But maintenance errors can never be eliminated entirely. To reduce them to a minimum, Service HQs (not project teams) must make materiel and financial provision from the outset to ensure the best training possible is provided and sustained. If that training is insufficient to satisfy the manufacturer's servicing instructions, which are agreed and accepted by MoD before production begins, then there is a disconnect that must be resolved - by MoD. The person making this provision is required to have been a maintainer. Rampant privatisation and outsourcing means MoD has fewer staff with the necessary experience, and the obligation has been almost impossible to meet since the early 1990s.

Often, associated risks are not recognised. As we will see, two of the three main risks that occurred on 8 November 2011 were only entered in the Red Arrows Risk Register after the accident. Yet they were standing risks. The system was reactive, not proactive. This was not an error of omission. It was one consequence of MoD's policy of *savings at the expense of safety*, whereby risks (and certainties) are allowed to occur before taking action.

*

Risks must be continually assessed, even those apparently mitigated. If maintainers arrive at the squadron without the necessary skills, or posts are cut, then the Risk Score (the product of probability of occurrence

---

7 'Breaking the Military Covenant - Who Speaks for the Dead?' (David Hill, 2018).

and impact) must be reassessed and the mitigation plan revalidated. Here, MoD had cancelled its training contract with Martin-Baker, ceased ejection seat training for armourers, abolished maintainer posts, and instructed those who remained not to implement the company's servicing instructions. The primary defences were consciously dismantled, making it more likely that, should the seat be needed, it would not work. The Risk to Life increased exponentially. Martin-Baker played no part in this.

A feature of this case is how MoD project teams are stove-piped, and the difficulty this causes in identifying and managing boundaries of responsibility. For example, parachutes and seats are procured and managed by separate teams, who create their own Hazard Logs. The 2015 'Safety Assessment Report for Emergency Escape Parachutes' lists as its top hazard *'Parachute fails to deploy resulting in unimpeded fall from height'*. The risk mitigation plan, correctly, requires *'adequate training and supervision'*. But, it claims this has been *'applied'* and therefore the risk is ALARP. Plainly, this is wrong, MoD itself admitting that training is inadequate, with directives issued to bypass critical procedures designed to ensure the parachute opens. Therefore, this particular Hazard Log entry is invalid, in turn contaminating the seat and aircraft equivalents.

The ability to *'deploy'* is the responsibility of the seat team; not the parachute team, who should be concerned with the ability of the parachute to *develop* after it has deployed. It can be seen that each team contributes to the correct operation of the escape system (as does the aircraft team), and all need to be aware of how *their* item relates to the others. As the interface between the parachute and seat teams' equipment is respectively the Drogue and Scissor Shackles, then by definition both have formally declared that they fully understand this interface. Otherwise, use of the equipment is strictly prohibited. The Health and Safety Executive's case was that no-one in MoD understood this interface and relationship, and that this was Martin-Baker's fault.

Whether or not these formal declarations were made, this mandated obligation is where investigators should have looked. This is an MoD organisational issue. That is not to say the organisational structure is wrong. But having chosen that structure, it is for MoD to ensure it is managed correctly. Plainly it was not, the Service Inquiry highlighting many associated failings; not least in the provision of Safety Cases. Martin-Baker played no part in this.

It should also be noted that, when the Mk10 seat was accepted by MoD, very different arrangements existed for equipment certification. Prior to the early 1990s, it was routine for MoD to enter into an arrangement to provide specialist facilities, meaning (in this case) the Royal Aircraft Establishment Bedford and the Aeroplane & Armament Experimental Establishment, Boscombe Down, played a major part. Policy later changed to:

> *'A contractor should have available all the necessary expertise to enable certification of the design to be carried out without a particular aspect of certification having to be made the subject of special arrangements between the contractor and MoD'.*

It will be apparent that this is not a simple case of Martin-Baker dealing exclusively with the seat project office, to whom they were under contract. There were MoD intermediaries who had to confirm they understood the design. To facilitate this, they were provided with full supporting documentation, operating instructions, and maintenance procedures. They were obliged to inform the project manager of any deficiencies; and he was obliged to seek clarification before signing the Certificate of Design.

*

In these circumstances, or if no longer under proper contract, how quickly and to what extent do a company's obligations recede? How long does it wait without income before making staff redundant? If or when a contract reappears, can former staff be re-employed? Do they want to come back? If not, how long does it take to recruit and train new staff? Major contractors with a better cash flow than Martin-Baker faced this dilemma after Options for Change (1990), warning MoD of huge start-up costs should it ever wish to resurrect capability or safety management. A spending bow wave became inevitable. And behind it came the trough.

5: Safety

The Safety Case

The decision to conduct Service regulated flying must be supported by a Safety Case. For aircraft, MoD defines it as:

'The study of an aircraft or item of aircraft equipment to identify and show acceptability [or otherwise] of the potential hazards associated with it. The Safety Case provides a reasoned argument supported by evidence, establishing why the Design Authority is satisfied that the aircraft is safe to use and fit for its intended purpose'.

There are other definitions, but this one is excellent because it mentions both aircraft and aircraft equipment, which MoD makes provision for and manages in separate ways - particularly relevant to this case. Also, it makes clear the Design Authority's responsibility.

A Safety Case considers five questions:

1. What are we looking at? (System description)

2. What could go wrong? (Hazard identification and analysis)

3. How bad could it be? (Risk assessment)

4. What has been or can be done about it? (Risk and ALARP appraisal, mitigation and acceptance)

5. What if it happens? (Emergency and contingency)

It should answer these for each of the uses defined in the Statement of Operating Intent and Usage, and for each Build Standard.

The Safety Case is a significant body of evidence. It is condensed in the Safety Case Report - a gradual refinement, condensing the work at a given time, setting out residual risks and updated at key points, such as when the form, fit, function or use changes. If there is no valid Report, or if the Build Standard is not maintained, the validity of the Safety Case erodes. If the Safety Case is not reconcilable with the rest of the Aircraft Document Set, flying becomes increasingly difficult to justify.

The degree of invalidity varies depending on (e.g.) how up-to-date the supporting evidence is, and is a matter of professional judgment. Plainly, therefore, if the professional expertise that underpins the Safety Case is removed, by cutbacks or natural wastage, and not replaced, then the argument must be reinforced from another quarter. This is one of

MoD's greatest failings.

*

In 1990 the RAF's run-down of airworthiness management commenced in earnest. In January 1993 it finally withdrew all funding to maintain avionic Build Standards and Safety Cases/Arguments. (The domain I worked in at the time). By end-1993 most expertise had vanished, because if there is no money MoD and contractor staff can do nothing and their posts disappear. When partially restored, no retrospective action was permitted to plug gaps. One had to pick up the pieces and hope that no gaps proved fatal. A forlorn hope, 29 dead on Beinn na Lice, Mull of Kintyre, on 2 June 1994 immediately proving the point.

The mandate for Safety Cases only flowed down to Martin-Baker in 2001, through a requirement for energy-absorbent seat cushions. Hitherto, MoD's ejection seat offices had been content for aircraft offices to produce a Whole Aircraft Safety Case, simply assuming the seat was safe. Prior to 1992 the system was less formal, older equipment often enjoying 'grandfather rights' whereby it was deemed to be safe and exempt from continual changes in legislation. But that did not mean one could ignore safety, and these assumptions were still subject to continual assessment. In 2001 someone acted properly by requiring a Safety Case; but it seems he was the only one, the team reverting to previous practice once he left. (Ladies, it was a *he*. It was a *she* who managed the task at Martin-Baker). The Service Inquiry reported this state of turmoil remained in 2012.

*

Design of the 70+ Mk10 ejection seat variants is controlled under configuration management procedures. The Aircraft Design Authority subsumes the seat Safety Cases within the Whole Aircraft Safety Case. This requirement to provide subsidiary Safety Cases was set out in inter-MoD project team Service Level Agreements, but they were not enforced. By 1996 few teams would discuss them, never mind sign one. Reaction ranged from refusal, to incredulity that Safety Cases still existed. Many said: *Stopped doing that years ago when the RAF pulled the plug*. Most agreed the work was necessary, but it was no longer policy because a policy remains an aspiration until resourced. However, it is one thing not to have the wherewithal, but quite another to be told to ignore safety and make false declarations that legal obligations have been met.

Here, the event requiring a recent seat Safety Case Report was Top Cross Beam cracking. (Explained later). Additionally, an updated Whole Aircraft Safety Case, and the higher level Air System Safety Case, was required, because only they provide operating context and address on-aircraft maintenance. Neither was produced.

(After initial publication in 2018, current and former RAF staff sought me out. As of August 2021, there is still no Hawk Air System Safety Case. It cannot be created as historical records (the audit trail) are damaged, missing or destroyed. Is this related to the recent announcement that Hawk T.1 is to be withdrawn from service by March 2022? This follows quickly on the heels of the withdrawal of Air Cadet glider fleets for the same reason - MoD could not demonstrate they were airworthy. Both aircraft types are managed by the same teams in MoD and the RAF. Yet the Red Arrows are to retain the Hawk T.1. One can only assume this explains the panic to create a Safety Case. But under what authority is the aircraft currently flying? Is accepting such a risk after so many airworthiness related fatal accidents a gamble worth taking? And what action has been taken about the decades of false declarations that there *was* a valid Safety Case? Are these people, or their acolytes, still running the show?)

*

Post Design Services (PDS), defined as 'maintaining the Build Standard', provides the contractual vehicle, oversight and control for maintaining Safety Cases. That fact alone automatically makes PDS the most important function of in-service support. There are 17 distinct core components, which can be viewed as a roof of interlocking tiles. Remove one and the rain gets in, gradually rotting the structure, and eventually the foundations. Outwardly all looks well, the missing tile hard to spot from ground level. The key is flexibility. One ensures this by implementing PDS Specification 18 when structuring the contract pricing.[8]

But the Military Aviation Authority, formed in 2010 after the Nimrod Review, uses an incorrect definition of PDS. Any project heeding it would soon find itself short of funding and unable to maintain a valid Safety Case. So, to do the job properly staff must ignore their regulatory authority - a weak foundation for any Safety Management System. The

---

[8] Defence Standard 05-125/2, PDS Specification 18. ('Requirements for Firm-Price and Actual Hours at a Firm-Rate PDS Contracts').

Nimrod Review confirmed this part of the airworthiness chain had broken down throughout MoD, a result of conscious decisions.

So important is this to understanding the accident, it is worth reproducing the Service Inquiry's comments:

> 'Regulatory Article 1220 states: "The Project Team <u>shall</u> produce and update a Safety Case". Neither the UK Military Flying Training System Team nor the Airborne Escape & Survival Team was able to produce a Mk10 ejection seat Safety Case Report, the latter believing it was held by the Hawk Support Authority. [Project teams exhibited] wider airworthiness shortfalls [and] weak audit trails to justify decisions, raising questions over the demonstrable rigour being applied to airworthiness decision making. This <u>general shortfall in auditable airworthiness decisions</u> fundamentally undermines Safety Cases'.

Numerous staff were required to certify these reports existed and were valid. Post-Nimrod Review, this failure is nothing less than gross negligence and maladministration. The process failed, catastrophically, despite the Military Aviation Authority being created to prevent recurrence. Why, after so many deaths, is this still happening? This is why the report, and the prosecution of Martin-Baker, is so alarming.

Notably, the thoughts of the Ordnance Board (now part of the Defence Safety Authority) are not recorded. It was required to contribute to the Safety Case Report.[9] But if there isn't one, what can it do? The answer is - escalate. But that would be to complain about their superiors, emphasising the latter's lack of independence.

*

When a pilot signs for an aircraft, he should be able to assume the equipment he is strapping into has been thoroughly analysed, tested, certified and maintained. That, if anyone in his command and engineering chains has changed the configuration or servicing method, then that change has been fully analysed and checked against the current Safety Case. The Safety Case Report and Release to Service provide this assurance. On Hawk, on 8 November 2011, it was false. The MoD and RAF had done none of this work.

The Assistant Chief of the Air Staff signs to say Build Standards and Safety Cases are maintained. At a certain level people just sign papers thrust under their noses. But when putting my name to a binding document saying I have assured myself there is a valid Safety Case (and

---

9 Joint Service Publication 520, Part 2, Volume 7, paragraph 2.

I have, many times), I expected the latest Safety Case Report or Statement to be attached. Here, the inability to produce one would automatically raise a red flag. What on earth was said at monthly reviews? We don't know, because the Military Aviation Authority won't go there. Why? Because junior staff have already been, and their warnings ignored. To go again would reveal the truth of the lie. But we do know there was no MoD Ejection Seat Safety Case prior to 2001. And whatever was produced in 2001 quickly became invalid. In fact, it was seemingly discarded, as the Service Inquiry drew a blank in 2012.

Past Assistant Chiefs of the Air Staff need to explain how they came to issue a Hawk Release to Service. There is only one way it could fly in Service - formal acceptance that known Risks to Life could be reasonably eliminated, and a written explanation why they had not been. Making such a statement on an ejection seat would be insane. However, it must be said the RAF has a history of such folly; witness the signing of the Chinook HC Mk2 Release to Service in 1993 despite being under mandate not to.[10]

The seat and aircraft project teams had passed recent audit. The Service Inquiry report mentions this but stops short of criticism. It omits it has been past policy, and current practice in many areas, not to maintain Safety Cases. The Military Aviation Authority, and even Defence Ministers, were aware of this before the accident, and condone it.[11]

*

In July 2014, MoD invited bids to prepare Safety Case Reports for Aircrew Escape and Survival equipment. The reason given was:

> 'Documentation does not satisfy current requirements, exposing Duty Holders and potentially users to unknown or unquantified risks'.[12]

'Current' implied past mandates were satisfied. They were not. Embarrassingly, MoD listed 62 items lacking the requisite Safety Cases or certification. For example, the GQ5000 main parachute used in Tornado Mk10A seats. This, 12 years after an independent safety assurance report to the Tornado and ejection seat project teams warned that the GQ5000 was not certified and could not be used with the

---

10 Mull of Kintyre Review report, paragraph 2.2.8.
11 For example, meeting with Minister for the Armed Forces (Nick Harvey MP), 17 January 2011, attended by the Military Aviation Authority.
12 Tender FTSSTS/AC/7104, 22 July 2014.

Mk10A seat in Tornado.[13]

This is not about Martin-Baker. It is about MoD not being able to produce evidence to justify flying the Hawk. Not for the first time, it is placed in a perilous legal position. Or would be, if legal authorities regarded the reckless making of false record an offence.

Accountability and responsibility

On 29 November 2016, Air Marshal (later Sir) Richard Garwood, Director General Defence Safety Authority, gave evidence to the Defence Select Committee. While discussing Duty of Care:

'If we had a fatality in the military tomorrow, I could give you the four names for any part of military defence who have accepted personal accountability for that. Perhaps I could refer to one of our Duty Holder letters from the Chief of the Air Staff, Sir Andrew Pulford to Air Vice Marshal Turner, who is an Operating Duty Holder. It says: "You are personally legally responsible and accountable through the Secretary of State for air safety, the air systems and functional safety in your area of responsibility". We are now crystal clear <u>in the military</u> about where that accountability lies and it is not at lower levels, but at pretty senior levels: Lieutenant Colonel up to Chief of the Air Staff in this instance, and above to the Secretary of State'.

He said *'in the military'* because this has always been clear to civilian staff. But these Duty Holders do not have the authority to ensure they are able to meet their notional obligations. Take, for example, the consistent ruling that functional safety can be waived, yet a false declaration made to aircrew that it has been achieved. A Duty Holder might rail against it (none have), but he would need Ministerial support to do anything (as Ministers have upheld the rulings). And it's not just about *'senior levels'*. It's about everyone being prepared to challenge maladministration and illegal orders. And while I can't speak for any Lieutenant Colonels, I do know how their civilian counterparts are treated if *they* report airworthiness failings. One is left in no doubt that a career brief on your brief career will follow if you say another word.

MoD's Regulatory Article 1210 states that in the event of a fatality the Operating Duty Holder (here, Air Officer Commanding 22 Group) is expected to defend his tolerable and ALARP statement in court. Where

---

13 QinetiQ report AT&E/CR00782/1 'Mk10A Ejection Seat Modifications (02097 & 02198) for Tornado GR4/4A and F3 Aircraft - Phase 2', paragraph 1.4.21, December 2002.

was he to explain the decision to void previous certification and fit an unserviceable seat to XX177? This is the big question in this case, not the alleged lack of servicing information. RA1210 is exposed as a toothless sham.

Blood on their hands

The following senior staff and organisations were notified of the failings. Primarily, the predicted effects of the *savings at the expense of safety* policy. Only those marked # accepted the truth and took any action; albeit ineffective in the face of very senior resistance. With two exceptions, the individual posts are at 1-Star level and above. (Senior RN Captain/Commodore, Brigadier, Air Commodore, and civilian equivalents).[14]

- Equipment Accounting Centre (1988-1991) #
- Flag Officer Naval Air Command (1991-92) #
- Director General Support Management (RAF) (1992) (The senior RAF officer who, in December 1992, threatened civilian staff with dismissal for refusing to commit fraud).
- Director Military Aircraft Projects (1992)
- Deputy Director Support Management 11 (Avionics) (RAF) (1992-93) # (The most junior civil servant listed, line manager of those threatened with dismissal in 1992).
- Director Support Management 3 (RAF) (1992-3)
- Director Internal Audit (MoD) (1993-96) # (Author of the June 1996 report to the Permanent Under-Secretary of State confirming conscious waste, and supporting those threatened with dismissal).
- Director Procurement Policy (Project Management) (1994)
- Management and Service Organisation (1994-96)
- Permanent Under-Secretary of State for Defence (1996-2006)
- Director Helicopter Projects (1996-99)
- Chief of Defence Procurement (1998-99)
- Director General Air Systems 2 / Executive Director 1 (1998-2001) (The civilian 2-Star charged with management oversight of e.g.

---

14 Before 1997 Commodore RN was an appointment, not a rank.

Nimrod and Chinook programmes).
- Director Helicopter Support (1999-2000)
- Director of Personnel, Resources and Development, Defence Procurement Agency (1999-2003)
- Deputy Chief Executive, Defence Procurement Agency (2000)
- Director Support Operations (Rotary Wing) (2000)
- Various Ministers for the Armed Forces (2003-on)
- The House of Commons Defence Select Committee (2004-on)
- The Public Accounts Committee (2004-on)

Post-Nimrod Review notifications:
- The RAF Provost Marshal (2009-11) (A Group Captain, the most junior military officer listed).
- The Mull of Kintyre Review (2010-11) #
- The Chief Constable, Thames Valley Police (2010-13)
- The Secretary of State for Defence (2011)
- The Civil Service Commissioners (2012)
- The Independent Police Complaints Commission (2012)
- Heads of the Civil Service (2012-on)
- The Parliamentary Standards Commissioner (2013)
- The Prime Minister (2013)
- The Health and Safety Executive (2016-on, in this case)

I thought it important to list these dates and posts. Despite knowing the truth, the 2009 Nimrod Review dated the failings to 1998. The effect was to protect those responsible. MoD exploited this by continuing to denigrate its staff who had originally reported these violations, and who correctly predicted the outcome.

Most of the above are (were) senior civil servants. The Civil Service Code requires that they conduct themselves with integrity, honesty, impartiality and objectivity. Did they? No, emphatically not.

6: Accident causation & defences in depth

Most accidents can be traced to one or more of four levels of failure. The normal method of describing this is the Reason Model of accident causation, after Professor James Reason; more commonly called the 'Swiss Cheese' model or cumulative act effect.

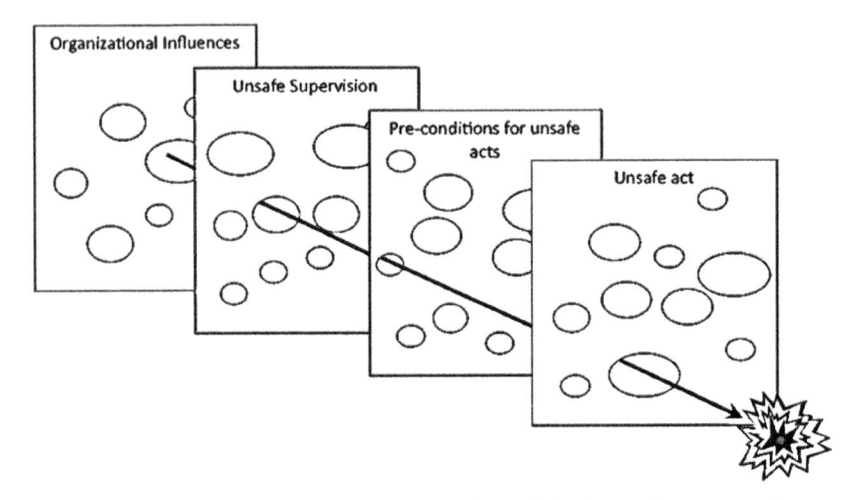

Figure 6 - The Reason 'Swiss Cheese' Model of accident causation.
(Professor James Reason)

Defences against failure are modelled as layered barriers, represented as slices of Swiss cheese. The holes in the slices denote individual weaknesses in individual parts of the system, and are continually varying in size. The system as a whole produces failures when a hole in each slice momentarily align, permitting the 'accident trajectory'; so a hazard passes through the holes in the defences, leading to a failure.

The hazards/failures may exist all the time, but not every aircraft will crash because the holes are seen to be constantly moving and only rarely, if ever, align. The model includes active and latent failures. The latter is useful in accident investigation, encouraging the study of factors that may have lain dormant, often for years, until they finally contribute to the accident.

Each level is divided into categories (with examples):

- Unsafe Acts: errors and violations. Errors are unintentional behaviours, while violations are wilful disregard of regulations.
- Preconditions for Unsafe Acts: condition of operators and personnel, environmental factors.
- Unsafe Supervision: failure to correct known problems.
- Organisational Influences: The broader, often indirect and latent, influences that a higher organisation brings to bear on those involved in an occurrence, and beyond the control of individuals; such as organisational climate, processes and resource management.

Inadequate defences may make errors more dangerous, but some errors will overcome even the most robust defences. In practice, defences should be strengthened in recognition errors <u>will</u> occur.

The model assumes several random events occur in such a way the holes align. But what if they are aligned by conscious acts such as denial of resources, or directives to make false record? When this occurs it is no longer an example of the Reason Model since it is not random - it is gambling. MoD is a chronic gambler. For example, for 17 years it blamed the two pilots for the Mull of Kintyre accident in 1994 (Chinook ZD576, 29 killed), having gambled on the worst failure possible in aviation - the aircraft was not airworthy.

*

I should expand a little on 'violations'. Matters are not black and white. In many cases the 'system' is so complex and contradictory that it sets you up to fail. Poor procedures. Time pressure. Equipment shortages. Inadequate training. Few staff. Poor leadership. Correctly, MoD acknowledges that risks may sometimes be disregarded and rules broken - often essential to achieving the military aim. But it differentiates between exceptional (acceptable) rule-breaking, sabotage, and doing so for personal gain, in a process known as FAIR (Flowchart Analysis of Investigation Results). Whether FAIR is administered fairly is a matter of debate. Junior officers and NCOs may make a correct assessment, but escalation is fraught with danger.

Most violations I cite arose from the *savings at the expense of safety* policy of 1987. This started out with a noble aim - to make efficiency savings while maintaining operational capability and effectiveness. However, its implementation was deeply flawed, immediately creating Risks to

Life.[15]

This was an easily fixed error. The officer who drafted the policy, a Wing Commander, agreed. It was to be used alongside extant regulations. Instead, his Supply colleagues cast both aside. It became a mistake when this was immediately pointed out, but his superiors refused to correct it. It became a violation when disciplinary action was taken against civilian staff who refused to obey orders to waste money and make false record.

The inevitable impact was notified in January 1988.[16] Military effectiveness and safety were being severely compromised. Sabotage? No, but it was certainly a conscious decision. There was much acclaim to be had for 'saving' money, but short-term postings meant others would have to deal with the fall-out. I justify this by pointing to the inordinate level of resources MoD continues to throw at protecting these senior staff, while vilifying those who sought to avoid the waste and keep aircrew safe.

These staff did not arrive at work intending harm to aircrew. But they certainly arrived intent on bullying and harassing staff into making false declarations. I could cite many examples, but need only point to the Cabinet Secretary's ruling of October 2014, that it remains an offence to refuse to obey such illegal orders.[17]

How an accident trajectory may develop

Since 1990 MoD has permitted the appointment of non-engineers to technical supervisory posts, enlarging a hole in the **Organisational Influences** slice. If they self-delegate airworthiness authority, the **Unsafe Supervision** slice disappears. If they further abuse that position and overrule engineering design decisions, new and bigger **Precondition** holes are created. This violation of legal obligations is a significant **Unsafe Act**. Despite formal warnings as to consequences, MoD management encouraged all this and the **Organisational Influences** slice crumbled altogether.

These layered defences were systematically attacked on multiple fronts, resulting in a series of fatal accidents. Hercules XV179, Nimrod XV230,

---

15 Policy initially advised to Industry in letter D/DDSS11(RAF)/48/9, 30 November 1987, to the Assistant Director of the Society of British Aerospace Companies.
16 Memorandum D/ARad 130/19/17 AWL32B/332, 14 January 1988.
17 Unreferenced letter from Cabinet Secretary Sir Jeremy Heywood, 28 October 2014.

Tornado ZG710, and many more. What would have prevented them all? Implementation of mandated airworthiness regulations. The loss of Sean Cunningham was just the latest of these deaths.

Hence, singling out Martin-Baker, even if they had committed the alleged offence, was never going to get to the root causes. That discrimination and pursuit, by the Health and Safety Executive, MoD and legal authorities acting in concert, did not get at the truth. It diverted attention away from it.

PART 3 - THE ACCIDENT, MOD INVESTIGATION, AND TECHNICAL MATTERS

'LETHAL WARNING: On entering the cockpit/cabin, it is the responsibility of the individual to ascertain that the correct positioning of the assisted escape system safety devices in each aircraft type and Mark are as detailed in the safety and servicing notes and Aircrew Manual related to aircraft'.

Warning in MoD Air Publications

(In Air Publications, a Warning implies the possibility of death or injury; 'lethal' is merely emphasis. A Caution implies the possibility of damage to the aircraft or its equipment).

'Supervisors take note: It has been said that nothing is fool proof because fools are so ingenious. Personal safety for those who work around ejection seats cannot be guaranteed; however, a high level of safety can be achieved if personnel have the proper attitude, understanding, training, and supervision. Unless proper maintenance procedures are followed exactly, even routine ejection seat maintenance tasks can bring about an accident or injury. Education of the workers is the best assurance for personnel safety'.

Warning in US Navy Aviation Structural Mechanic Training Manual

'This isn't something that has crept in over the past couple of years - I suspect it goes much further back in time - and whatever the structure of the Air Force there has been a chain of command at the top, monitoring and allowing the team to actually perform. I cannot believe this has all come as a big shock to them and it's hard to understand why they were not doing something more active to change that drift'.

Retired fighter pilot Air Vice Marshal Jerry Connelly, speaking after the 2014 Inquest.

7: The Accident

An accident is an occurrence resulting in a person being killed or suffering a major injury. It is the realisation of a hazard becoming a harmful outcome. An incident is an Air Safety related occurrence not resulting in an accident, but resulting in a person receiving a reportable over three-day injury, or an event compromising Air Safety.

Notably, if <u>any</u> cause of the entire sequence of events is already known, then an accident has occurred. Thus, even if Sean had survived without injury, this was an accident because Risks to Life were known but not tolerable and As Low As Reasonably Practicable.

The clues as to areas of concern came thick and fast. Chief among them, the very next day retired Air Chief Marshal Sir Michael Graydon, former Chief of the Air Staff, said this to the BBC:

'The Red Arrows will come through this and I think they will continue to use well proven methods of selection and well proven methods of training'.[18]

Unfortunately, the BBC has a habit of selecting unsuitable persons to interview as 'impartial' analysts, and not making their background or personal (or conflicts of) interest clear to its audience. Perhaps uppermost in Graydon's mind was the legal test for Corporate Manslaughter, which includes *'wider considerations such as...the selection and training of staff'.*[19] The Service Inquiry later severely criticised both. The choice of words might be thought too much of a coincidence, given the same issues, and his own culpability, had arisen during the Mull of Kintyre case.

The two final acts

1. <u>The inadvertent ejection</u>

When making the seat safe after flight, the Seat Pan Firing Handle is locked by one of six safety pins inserted by the aircrew. They are marked: Seat Firing, Main Gun Sear, Rocket Initiator Sear, Manual Separation Sear, Canopy MDC (Miniature Detonation Cord) Firing Handle, and MDC Firing Unit. They must <u>visually</u> check the Handle is fully seated in its housing and the Seat Firing Pin properly fitted. Rather

---

18 www.bbc.co.uk/news/mobile/uk-england-15653484
19 https://www.cps.gov.uk/legal-guidance/corporate-manslaughter

obviously, one does not pull the handle to check. The pin should not be re-inserted until the aircraft is stationary on chocks, with the engine(s) shut down.

The Aircrew Manual carries two warnings:

1. 'When the Seat Firing Handle safety pins, the MDC firing handle safety pins and the MDC firing unit safety pins are correctly fitted to their respective units in each cockpit the aircraft is Safe for Parking'.

2. 'When all (six) pins are fitted to their respective units in each cockpit the aircraft is Safe for Maintenance'.

The Service Inquiry noted aircrew were re-inserting safety pins as part of the after-landing checks, before the aircraft was stationary. That said, the Red Arrows have less capacity for scanning the Handle when fitting the pin when taxiing in formation, hence why it may be done during the landing rollout. Nevertheless, it is a departure from laid down procedures, so requires a waiver and an associated update to the Safety Case. But there wasn't one.

The Handle in XX177 had not been seated correctly, meaning the pin was not located properly. It remained in this unsafe condition for some days, unnoticed during 19 inspections by seven different people. On the day of the accident it finally dislodged, releasing the gun sear and initiating ejection. It was thought Sean had inadvertently misrouted one of his crotch straps through the Handle during the previous sortie; and that rudder pedal inputs made during *that* sortie had displaced the Handle, where it remained until crew-in for the accident sortie. Aircrew are instructed:

'Draw the strap between the legs ensuring it lies to the REAR of, and not THROUGH, the Seat Firing Handle'.[20]

The Aircrew Manual also warns:

'It is possible, when strapping in, to route the negative-g restraint [a separate strap secured to the front of the Seat Pan] *to the side and in front of the seat firing handle. If this occurs, the restraint can dislodge the seat firing handle from its housing when the seat firing pin is removed or the seat occupant moves. Occupants should make a positive check of the negative-g restraint routing before removal of the seat firing pin'.*

'When strapping in ensure that there is no possibility of the Aircrew Equipment

---

20 For example, AP108B-0122-1.

Assemblies (AEA) fouling seat mechanisms during subsequent ground operations and flight. A particular hazard has been identified with AEA snagging the manual separation handle causing initiation of the system'.

The Service Inquiry determined that leg movements while checking rudder operation placed tension on the strap and could dislodge an incorrectly fitted Handle.

*

How the pin can be incorrectly fitted is demonstrated in mandatory safety lectures for all personnel who enter a cockpit with a seat installed. One Survival Equipment lecturer:

'I delivered many 6-monthly seat lectures where I demonstrated incorrect strap routing and the danger of not inserting the safety pin properly. I know my colleagues did the same. Often during the 30-minute session I witnessed aircrew eyes glazing over and I sympathised; these were intelligent men who were quite familiar with the seat and resented being dragged away from their primary duties. I was in favour of extending the time between lectures to 12 months, but it looks as if regular warnings are still needed'.[21]

The procedures are easy to demonstrate on the classroom training seat, but a different matter when sitting in the aircraft. A factor here was the position of the Seat Pan Actuator (providing six inches of seat height adjustment). The battery switches in the front cockpit need to be switched to ON to raise or lower the seat. The Flight Servicing Manual requires it be motored to the upper limit to improve access for the loose article check, and offer a better view of the Handle.[22] But Red Arrows groundcrew had been told to leave it in position, the Panel concluding this was to save time during strapping-in; the perception being the Red Arrows wanted to be seen to do everything faster (but not necessarily better or more safely). Also, raising the seat fully shortens the leg restraint lines, so they must be extended again after the seat has been lowered. This appeared to be a fleet wide practice and not just at the Red Arrows.

So, while a 'better' way of operating had evolved, why was the Manual not amended? The process is simple. One raises an Unsatisfactory Feature Report on the Publications Authority, who tasks the Design Authority and MoD's specialists (for example, the RAF Handling Squadron) to come to a decision. That is, retain the status quo, change

---

21 E-mail to author, 15 March 2018 19:56.
22 For example, AP109B-0131-5F, Item 19.3.

the Manual, or grant a waiver to certain users. Either way, the Safety Case had to be updated. None of this took place, yet the Red Arrows had recently passed their Air Officer Commanding's Inspection with flying colours. How?

The short timeline that the Red Arrows practiced from starting briefing to take-off was a contributory factor, and is covered in the Service Inquiry report. The ejection seat checks would have been rushed and, combined with poor visibility/access caused by the seat being down, the probability of noticing the incorrect Handle position would reduce. This was verified during the investigation. Using the Red Arrows timeline the test subject missed the error, but spotted it when given more time.

Again, this goes back to what I said earlier. The Red Arrows must stand out because they are the best, not because they do things differently or more quickly. The recommended time to undertake these safety checks is determined through long experience. Doing things more quickly, and omitting things to finish more quickly, are two entirely different things. (And one could reasonably argue that a fixed time should not be applied to safety checks).

Nevertheless, had these been the only failures Sean would likely have survived. What killed him was multiple instances of incorrect application of airworthiness regulations...

2. The failure of the main parachute to deploy

The over-tight Drogue Shackle nut prevented the main parachute from deploying. The Service Inquiry said *'evidence suggested that MoD was unaware of the potential failure mechanism'*, saying there was *'no record of MoD receiving a warning'*.[23] It did not say who it asked or where it looked, but readily accessible evidence proved this entirely wrong.

And therein lies the main investigative and procedural error. To reach a point where it claimed no warning had been received, the Service Inquiry Panel must have (or should have) worked its way backwards through the process to the point where it *thought* it should have been issued. In doing so, it would have encountered evidence that it *had* been issued and received - but directives issued not to use it.

Also, numerous and long-term violations and breaches of duty, by MoD

---

23 Service Inquiry, paragraph 1.4.4.2.

personnel, would have stared it in the face at every turn. Yet it did not mention these, raising the suspicion that it did not look, desperate to avoid MoD liability.

At the Coroner's Inquest in January 2014 this morphed into a claim that, from the introduction of the parachute release mechanism design, MoD (and its predecessors) never had the information. The court did not hear evidence from those in MoD who had allegedly searched for the information, or from those whose job it was to manage the information (the data controllers). Or if they had spoken to each other. This was the first indication that the investigation had started down a misguided path, and refused to turn back.

By speaking to the right people, I was able to swiftly uncover voluminous evidence revealing the truth. The information had been provided many times, commencing in the 1950s. Heeding it would have saved Sean. However, MoD had instructed its staff not to use it, issuing contrary instructions. This was not explained in the Service Inquiry report or at the Inquest, tainting all subsequent media reporting and legal proceedings. What follows, therefore, is the only investigation to take cognisance of the known facts.

Normalisation of deviance

This phenomenon, whereby a clearly unsafe practice comes to be considered normal if it does not immediately cause a catastrophe, is a recurring feature of this and many other MoD accidents. Here, the incubation period was long. It was equally long on Nimrod, with fuel leaks caused by incorrectly fitted couplings tolerated for decades. And on Hercules, with the lack of Explosion Suppressant Foam the subject of long-standing aircrew concerns. In a way, one can understand this, as warning signs may be misinterpreted, or missed completely, and there is the perpetual pressure to maintain operational readiness. But look one level deeper and there is a common factor - the policy of *savings at the expense of safety*. There, the results were immediately fatal. The inevitable outcome had been predicted, notified, and ignored - and I listed in Chapter 5 some of those notified. Faced with that culture, and their abrogation of duty, normalisation of deviance became the norm.

8: The Problem

During a routine After Flight Servicing in July 2010, cracking was discovered in the front seat Top Cross Beam Block Assembly (to which the Scissor Shackle is attached) of Hawk T.1 XX263. It is closely associated with the main gun, which propels the seat out of the aircraft before the rockets fire.

The Service Inquiry noted that, historically, ejection seats and their guns were kept as matched pairs, but it *'could not be proven whether each seat is matched with its original ejection gun'*. It went no further, omitting that the same observation had been made by the Board of Inquiry into the loss of Tornado GR4 ZA554 on 14 November 2007 (see Chapter 24):

'Historically, ejection seats and guns were paired; however the process of matching was not conducted in the Seat Bay at the time of the final seat fits of ZA554. The Board believed that had the ejection seat been matched to its ejection gun...the fouling found by tradesmen may not have occurred'.[24][25]

Immediately, we have a recurrence caused by failure to implement previous recommendations.

On 28 July 2010 Urgent Technical Instruction/Hawk/026 was issued by the Hawk Support Authority at RAF Wyton, Cambridgeshire, to inspect the T.1, T1A and T1W fleets. It included this statement:

'This UTI is issued for data gathering purposes and a failure to identify cracking does not infer that the seat is, at this time, safe to use'.

What were users to make of this? The statement removed the safety verification assurance. That is, even if no cracks were found, Mk10B seats had to be considered unsafe until stated otherwise. The results were to be reported to British Aerospace under the Hawk support contract, managed at RAF Valley. Further cases were revealed. The Service Inquiry listed two, and the 1710 Naval Air Squadron Materials Integrity Group four. (1710NAS is the latest incarnation of what was the Naval Aircraft Materials Laboratory, MoD's material testing facility).

Martin-Baker, along with 1710NAS, conducted a test firing of a cracked seat. In November 2018 MoD finally released Service Inquiry Report

---

24 Tornado GR4 ZA554, Board of Inquiry report, paragraph 47.
25 https://apps.dtic.mil/dtic/tr/fulltext/u2/a248021.pdf Pages 5 & 24.

Exhibit 107, the 1710NAS report.[26] It revealed the cause was fatigue, a result of asymmetrical loading. The failure was of *'a progressive nature, and no single incident can be attributed to the cause'*. No adverse inference should be drawn. Fatigue cracking is common in aircraft structures, and the precise cause might be unavoidable forces acting on the seat. What is important is how it is monitored and controlled. Oddly, the report excluded information from the Red Arrows. Why?

Martin-Baker recommended a fortnightly visual (non-dismantling) examination.[27] Instead, the RAF elected to conduct Non-Destructive Testing crack detection every 50 flying hours - far in excess of the regular maintenance regime, but not as frequently as Martin-Baker recommended. 'Fortnightly' is significant. By definition that means it is *urgent*.[28] The 50-hour directive changed that to *routine*. Importantly, the proposed action must be complemented by resources, particularly relevant given the existing onerous maintenance burden. Crucially, it must be taken into account that frequently disturbing a safety critical aircraft system may, in itself, introduce (another) Risk to Life.

Development of Routine Technical Instruction/Hawk/059

The Hawk Support Authority, seat Engineering Authority, and the RAF's Non-Destructive Test Squadron, produced Routine Technical Instruction/Hawk/059. The Support Authority issued it on 2 August 2010, on behalf of Air Officer Commanding 22 Group. Please bear in mind that Martin-Baker, by definition, disagreed with this.

Urgent and Routine Technical Instructions (UTI/RTI) are Service Issued Instructions, meaning they are an internal mechanism. When used correctly they are expedient. But too often they are used to bypass the perceived red tape of airworthiness regulations, Design Authorities, and the correct people in MoD.

In July 2010 UTI/RTIs were governed by Joint Air Publication 100A-01, Chapter 10.5.5, which states a UTI must be satisfied within 14 days or 25 flying hours. An RTI is for less urgent actions. Both are issued:

'To inform users of a potential fault and audit the extent of a potential

---

26 Technical Memorandum MIG 10.376 'Investigation into the failure of a number of Mk10 fast jet ejection seat head top blocks', 14 March 2011.
27 Service Inquiry report, paragraph 1.4.4.4 (a).
28 Joint Air Publication 100A-01, Chapter 10.5.5.

problem'.

However, they are:

'Only to be raised when...there is no requirement for Design Organisation involvement'. ('Design Organisation' is erroneous - that is an accreditation. 'Design Authority' is the appointment being discussed).

Hence why they are internal only; and why the mandatory Defence Standard and Specifications setting out Design Authority duties do not mention them.[29] Patently, any work arising from cracking in an ejection seat requires the involvement of both Martin-Baker and British Aerospace, as maintenance procedures may have to be revalidated.

Therefore, an RTI was prohibited. Issuing it disregarded Martin-Baker's advice that the matter should be treated as urgent (the fortnightly inspection), and ensured no Safety Case assessment would be conducted.

The correct procedure was to issue a Special Instruction (Technical), which must always be appraised by the RAF Handling Squadron. There is no such thing as a *routine* Special Instruction (Technical) - they are all urgent.[30] For this reason, and because they do not ensure the Safety Case work is done, UTIs and RTIs are not Special Instructions (Technical). This was not understood within MoD, revealing a systemic failing.

*

There are various types of Special Instructions (Technical), issued by different MoD authorities. In the circumstances, and given (a) the parachute release mechanism was to be dismantled, (b) the application was recurring, and (c) Design Authority involvement was required, a Servicing Instruction was the correct route. These are issued to detail:

'Repetitive work to be carried out within a specified time or other limit, to seek to repair, or prevent, a potential fault. [A perfect description of the work required]. *Prior to issuing a Special Instruction (Technical), the MoD Project Team Leader will assess its effects on the aircraft or equipment Safety Case, Release to Service, handling and operation'.* [In practice this is delegated to specialists].[31]

The draft Instruction must be technically approved by the Design Authority. Anything thought to affect 'handling and operation' must

---

29 Defence Standard 05-125/2 and PDS Specifications 1-20.
30 Defence Standard 05-125/2, Chapter 7.13.2.
31 Defence Standard 05-123, Chapter 10.1

also be appraised by the RAF Handling Squadron. Only then is it authorised and issued by the MoD Technical Agency - not the Service. (The Technical Agency is the named individual responsible for maintaining the Build Standard, which includes maintaining the Safety Case). It would seem none of this took place, otherwise the lack of a Safety Case Report would have been noticed. (The RAF Handling Squadron has been partly replaced by the Defence Aircrew Publications Squadron, but it is unclear who is now responsible for the remainder of the work - which may explain much).

*

Upon application of RTI/Hawk/059, further instances of cracking were found. On 5 August 2010 it was up-issued to 059A to include additional Non-Destructive Testing. The work, scheduled to take 8.5 hours per aircraft (two seats), was to be carried out before the next flight. That is, the requirement had changed to *urgent* but was still covered by a *routine* order. The RTI did not stipulate if this was a recurring requirement.

On 20 August 2010 it was up-issued to 059B, clarifying the recurring periodicity. It was to be re-applied within 28 days from the last application of 059A or 059B.

On 17 September 2010 it was up-issued to 059C, further revising timings. It was now to be re-applied 50 flying hours from the last application of 059B, but could be deferred by 5 hours. (While a 10% extension is permitted in exceptional circumstances, I was always taught this would be unwise on a safety critical system, where the first failure could be fatal. This naturally raises the question - are other, longer, extensions being granted? A rhetorical question. 60% is now normal).

On 27 October 2010 it was up-issued to 059D, clarifying the actions necessary for seats not already fitted to aircraft. At this point someone should have noticed *these* seats would undergo a more extensive final test, and query the effect of not being able to do this *in situ*.

So many re-issues within two months was, in part, a product of not adhering to regulations requiring a Special Instruction (Technical); although some elements were, correctly, a product of feedback from inspections. However, the process was clearly mismanaged. A proper hazard analysis of the proposed servicing change was not conducted. This would have forced a review of the Safety Case, and an updated Safety Case Report - revealing (again) that there wasn't one.

*

Concerned over lack of information regarding the RTI, on 11 October 2012 the President wrote to the Aircraft Assisted Escape Systems Project Team in Bristol seeking *'further clarification'* and details of an audit trail. None existed. No record of how the RTI came to be applied to the seat. No entries in the Hazard Log. No record of key decisions, or why the RTI was not checked against Safety Cases. In signing the RTI, the Issuing Authority was confirming this audit trail was complete. In fact, he was only speculating, gambling with aircrew lives that all was well. He lost.

The Inquiry did, however, uncover that meetings had taken place *'to discuss operational and engineering aspects of the issue'*, chaired by a senior RAF officer from 22 Group (as the Aircraft Operating Authority). The attendees were the Technical Head of the Military Aviation Authority, the Hawk Support Authority, the seat Engineering Authority, 1710NAS, FLEET (who task 1710NAS and operate RN Hawks), and the Non-Destructive Testing Squadron. Together, they *'agreed the mitigation and way forward'*.[32]

These attendees perpetuated the primary risk, by agreeing *in situ* application without checking prerequisites were intact. And what was the role of the Military Aviation Authority (MAA) at the meeting? As regulatory authority it could not be responsible for both preparing the Safety Argument (which is part of the agreed mitigation) *and* declaring it adequate. I would not argue against its presence in an advisory capacity, but surely it should have advised the chairman that further serial breaches of mandated regulations were being proposed?

The Service Inquiry report does not say if the President challenged any attendees over these breaches. Perhaps he felt this outwith his remit. That it was for the MAA to pursue on an MoD-wide basis, given the same failings had been noted as systemic by the RAF Director of Flight Safety in a series of reports between 1992-98, the Nimrod XV230 Coroner in 2008, and the Nimrod Review in 2009. However, one would definitely expect them to be noted in the MAA's remarks. But the MAA was Convening Authority, regulatory authority, and an attendee, so would not be keen on highlighting this fatal conflict of interests. That being so, where was the independent regulatory/assurance oversight? The MAA's alleged independence is a distinction without a difference.

*

---

32 E-mail from Aircraft Assisted Escape Systems project team to Service Inquiry President, 18 October 2012 12:00.

The Inquiry confirmed Martin-Baker were not involved in the RTI.[33] MoD later claimed the company were *'made aware of and agreed with the RTI content'*; but then caveated this with *'within available MoD resources'*.[34] It is plainly wrong to say the company *'agreed'* with the RTI, as it went against the advice provided. They were presented with the RTI as a *fait accompli*. Knowing MoD was flouting its own regulations, they had a choice to make. Walk away, or try to make the best of a bad job. They chose the latter, as most companies placed in this position do.

'Available resources' clearly alludes to there being no Seat Bay at RAF Scampton. Also, insufficient spare seats to cope with them being removed and returned to the central bay for this unscheduled servicing.

What is lacking here, and indeed in any MoD report, is a company perspective on the impact of the RAF's *savings at the expense of safety* policy. I do not know how Martin-Baker reacted to this policy. But I have first-hand knowledge of what other major contractors, such as Ferranti Radar and GEC-Marconi, said in the late 80s/early 90s. It was not complimentary, and they predicted precisely what would happen. Vast waste, and a reduction in safety.

(Since initial publication it has been said to me, by a retired RAF engineer close to the investigation, that the company were *'not open'* to an investigation. I believe the foregoing provides context to this perception, and would almost certainly have been unknown to the MoD and police investigators - both of whom would be oblivious to the RTI being rogue. Whereas, if any Post Design Contractor did not know this, I would regard them as unfit for the appointment. For their part, British Aerospace were only involved for post-application reporting purposes).

Functional testing

The Drogue Shackle is part of the Duplex Drogue Assembly, and is pre-assembled (nut and bolt fitted) in a separate Survival Equipment Bay during preparation of the parachute attachment line prior to being packed. The Head Box is then brought to the Seat Bay and presented to the rest of the seat.[35] There, if the nut is too tight, this will be evident when carrying out the following checks, laid down in the Servicing

---

33 Service Inquiry report, paragraph 1.6.7.
34 DE&S letter FOI2017/08719, 22 September 2017.
35 AP109B-0132-5F Bay Servicing Schedule, paragraph 89.

Schedules. In strict sequence:

- ✓ The shackles are checked <u>twice</u> for disengagement, and re-linked.
- ✓ The Barostatic Time Release Unit is cocked, engaging the restraining plunger, preventing the Scissor Shackle from opening.
- ✓ The safety ties are fitted.

This sequence is impossible if attempted *in situ*, confirmed by Mrs Justice Carr in her remarks:

> *'Whilst it was possible, to a limited degree, to check whether there was free movement between the shackles, it was not possible to check whether the shackles could be released'.*

In effect, she pronounced Martin-Baker innocent, but did not understand the full implication of her words.

Disturbed Systems Testing

While the nut was, by any standards, fitted incorrectly, the main infraction was not verifying serviceability. The act of dismantling the Drogue Shackle invalidated the earlier signatures certifying correct function. Moreover, and to illustrate MoD's recent exposure to such matters, this precise point, about Disturbed Systems Testing, had been made in the main submission to the Mull of Kintyre Review in 2010.[36]

That is, if one has to disturb an existing system, for example to gain access to the equipment actually being serviced, then one must check that the disturbed system has been made serviceable again. The underlying principle here is that *servicing is not complete until verified*. Failure to do so had killed Red Arrows pilot Flight Lieutenant Simon Burgess, in 1996. (See Chapter 24). The Service Inquiry mentioned this accident, claiming it led to *'ground crew monitored flying control checks'*. But it omitted that the same root failure also killed Sean Cunningham.[37]

The cracking was underneath a Scissor Shackle arm, an area not covered by existing testing. Recognising the risk, this is why Martin-Baker had recommended a non-dismantling visual inspection. But to satisfy the Routine Technical Instruction the Scissor Shackle would have to be raised to the vertical, entailing cutting safety ties and disengaging the Drogue Shackle. Serviceability would have to be revalidated - the only

---

36 'The Inconvenient Truth' (David Hill, 2021).
37 Service Inquiry report, paragraph 1.4.5.19d

certain way being to check shackle disengagement. This is impossible *in situ*, as the Scissor Shackle is prevented from opening by the cocked Barostatic Time Release Unit.

One must not look at this in isolation. The inability to check an item will actually work is fairly common, a parachute being a good example. This is mitigated by careful design and extensive testing, throughout the design envelope, and comprehensive training and oversight. If instructions are followed to the letter, then it will work. But here, MoD, not Martin-Baker, forsook that option by ordering *in situ* dismantling of the Drogue Shackle by untrained maintainers, using the wrong tools, and contrary to the technical publications. A waiver was required, and many people of differing specialisms (e.g. Engineering, Quality, Risk, etc.) were required to sign it. No report mentions this. And it is all the more important here, because there is no allowable failure rate; reflected in Martin-Baker's instruction to check disengagement twice, in the Seat Bay.

Investigators should have observed while a tradesman read the Routine Technical Instruction, Martin-Baker manuals, and Air Publications, noting his reaction when attempting to implement them. He would find it impossible, which should have stopped the investigation in its tracks. At this point a decision as to liability would be possible. Martin-Baker would not be mentioned.

*

A by-product of the modular design of the Mk10 seat was that the Head Box could be removed and refitted *in situ*. But that does not mean one should adopt this as a maintenance policy. For example, the design of my PC allows me to remove the cover and take out the processor while it is running, but common sense says it will not work again. One might have to explain this principle, once, to a five-year old. But not to trained engineers. Here, the obvious stumbling block is that the disengagement check cannot be carried out when the Head Box is refitted.

The real, and intended, benefit of the design was to allow the Head Box to be serviced/assembled in a separate bay, by parachute specialists, making the packing easier. The intended *in situ* benefit of the modular design was to allow the seat pan to be *'rapidly removed to provide full access to the equipment in the cockpit without disarming or removing the seat main structure or aircraft canopy'*.[38] The Head Box redesign and re-positioning

---

38 Martin-Baker, leading particulars of Mk10 ejection seat.

was to *'permit rapid smooth deployment of the parachute and drogues, clear of the seat structure'*.[39]

Also, in some aircraft removal of a complete seat is expensive and time consuming, as much of the cockpit structure must be removed. A modular construction makes things easier, but in no way implies each module can be removed and serviced separately. In summary, the design intent was not to facilitate *in situ* Head Box changes, or dismantling of the Drogue Shackle. Quite the opposite.

*

The risk would be immediately apparent. The decision to accept it was an MoD decision, not Martin-Baker's. Given the safety criticality of the escape system, and the required waiving of safety mandates, the approval of the Senior Duty Holder was required - the Chief of the Air Staff - because <u>no-one</u> subordinate to him has the authority to accept such a risk. Yet the Routine Technical Instruction broke every rule in the book. It is unclear if those involved knew there *was* a book.

Still, one can discern the theory. If the instructions for assembling the Drogue Shackle nut and bolt, and then fitting it within the Scissor Shackle jaws, were followed, then the Shackles <u>should</u> disengage when designed to. But is *'should'* sufficient in a safety critical application? No, it is not. There are too many diverse dependencies outwith the control of any officer asked to agree such a deviation; including the Chief of the Air Staff. One could reasonably argue, therefore, this was a decision for the Secretary of State. The obvious problem is that the people who would advise him are those who were proposing the illegal actions.

The strength of the defensive barriers against failure must be known, and remain intact. As part of the decision-making process this would have to be <u>reverified</u> and a positive statement made. If this mandatory risk assessment was carried out, what did it say? If the necessary information was not available (as MoD and the Health and Safety Executive claim), why did the policy change proceed before it was sought, delivered and verified?

Which is it to be, MoD and HSE? You can't have it both ways.

*

In a legal sense, the signature on the Routine Technical Instruction is a firm statement that the signatory has satisfied himself that the necessary

---

39 Martin-Baker, Mk10 ejection seat brochure.

information is available. Whether he understood this is another matter, but would itself be a serious Organisational failure. Therefore, the charges against Martin-Baker were completely baseless, on a number of counts.

During the trial, in 2018, the Health and Safety Executive claimed all this was irrelevant. That, it was Martin-Baker's sole and *'non-delegable'* responsibility to ensure the parachute deployed and developed under any circumstance, regardless of malpractice or customer negligence. This has no legal basis. Yet the Judge did not demur. In fact, it was MoD's Duty of Care that was non-delegable.[40]

Thus, we can identify the primary root cause. The conscious decision to erode safety margins to a dangerous level, in the face of dire warnings from the Director of Flight Safety, meant untrained maintainers, given incorrect instructions in a rogue RTI, were directed to carry out the work *in situ*. There was no check the defences in depth stood, and there was no disengagement check. This situation arose from the RAF's policy of *savings at the expense of safety*.

*

Were it my job to sign off on waiving the disengagement check, I would not. (And for some years it *was* my job; not on seats, but on other safety critical equipment and components). Mainly because of lack of control over all the elements. I would look at other recent fatal accidents, such as Nimrod XV230 and Hercules XV179, where the same failure to assess cumulative risks occurred. Despite each being notified to the correct person, nothing was done. That failure to act is endemic; again taking responsibility outwith MoD, to the Secretary of State. No politician has shown the will to correct matters. With reference to the Reason Model, that is a full house. An Organisational Failure, Unsafe Supervision, a Pre-Condition for an Unsafe Act, and an Unsafe Act.

Premise

This book therefore proceeds on the basis that MoD's refusal to meet its obligations meant the risk of shackle pinching was no longer tolerable and ALARP, and so Disturbed Systems Testing should not have been waived. I accept there are many other similar dilemmas facing MoD

---

40 Woodland (Appellant) v Essex County Council (Respondent), 23 October 2013.

staff. But few relate to safety critical escape systems that are required to work perfectly, only once.

The circumstances of Sean's death justify my arguments. The defensive barriers didn't just fail. They were actively dismantled when the RAF refused to use Martin-Baker's instructions, ceased training, and allowed the incorrect tools to be used. MoD then rid itself of the necessary corporate knowledge. The probability of occurrence increased exponentially, rendering this an accident waiting to happen. In fact, I would go further. The sheer scale of the recklessness and maladministration involved, in the face of decades of warnings to very senior staff, requires consideration of manslaughter.

9: Applying the Routine Technical Instruction

Owing to their hazardous nature, any work on ejection seats must be authorised to personnel who have undertaken seat safety training. Supervisors are required to verify that those carrying out any work have the required qualifications, training, publications, and tools to do the job (although they have little control over any of this). Tradesmen must understand how the system they are working on functions and fails. They don't just fix problems, they must be able to diagnose them.

However, MoD's policy and practice had changed long before 2011, disregarding these fundamentals. Engineering training had been curtailed, posts cut, and inexperienced tradesmen with no authorised training permitted to work on the aircraft. They (and more senior engineers) were no longer required to understand *why* they were doing something. It was sufficient for them to be told, in technical publications, when, where and how.

The impact of such a proposed maintenance (and personnel) policy decision must be assessed, and compensatory measures put in place, before implementation. Often, publications must be completely re-written, as they assume a minimum level of training and competence. This is MoD's responsibility, not industry's.

Here, the technicians' lack of understanding as to how the seat worked meant they failed to appreciate the effect of over-tightening the nut. By the time they were applying the RTI to XX177 (on 24 October 2011), they had been let down by superiors. As too had Sean Cunningham.

*

No investigation asked if the correct information was with the correct people, in the correct place. Instead, the wrong people were asked about the wrong information. What documentation did the technicians have when working at the aircraft? Not Defence Standard 00-970 (Design and Airworthiness Requirements for Military Aircraft), as implied by MoD - a deceit which took on greater significance when the Health and Safety Executive repeated it and the Judge based her sentencing remarks on it. They had Air Publications and the Routine Technical Instruction.

Lacking comprehensive trade training, their engineering judgment was limited; leaving common sense, which is a poor basis for sole culpability. However, they were culpable to a degree because they should have recognised anomalies and known to challenge them.

But degree is important. The cultural pressure on young RAF technicians to 'obey the last order' is immense. To apply common sense would be seen as criticism of senior officers and the regulatory authority. Their line manager might praise them, but there would be a reluctance to escalate. This is precisely why MoD Technical Agencies and Design Authorities are charged with visiting Air Stations to have frank exchanges with aircrew and maintainers - part of the feedback and continuous review obligation. This is a core component of maintaining the Build Standard; which, we have seen, is a prerequisite to a valid Safety Case and Service regulated flying.

More profoundly, lacking proper training they had no right to be anywhere near an aircraft with tools in their hands.

The Drogue Shackle Assembly

Let us remind ourselves what they were working on, because failure to conduct the work properly led directly to Sean's death.

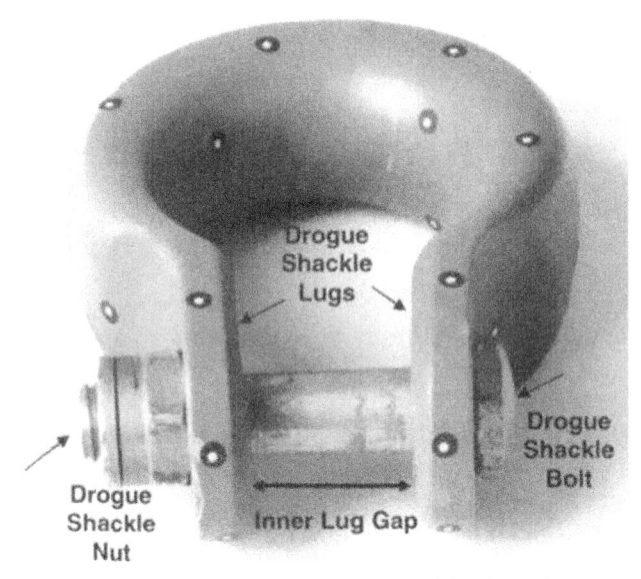

Figure 7 - XX177 Drogue Shackle. *(MoD)*

The nut is a Philidas stiffnut, with one or more slots cut laterally in the reduced-diameter circular top. Over the last one or two turns, the thread for at least half the diameter is elastically displaced, increasing the friction between nut and bolt, creating the locking action.

The Service Inquiry used the term 'locknut', which is different from a stiffnut although often used as a generic term. But in this application the distinction is important, because the regulations on re-use differ depending on type of nut. For example, all-metal stiffnuts (like the one shown) will act like a die and remove protective plating from both nut and bolt, damaging their threads. This wear and tear loosens the grip if either is re-used. This was a major factor in the loss of Sikorsky S-92A C-GZCH off Newfoundland on 12 March 2009, killing 17. The report into this accident was issued shortly before the XX177 accident, and hosted on the Health and Safety Executive's website. It is reasonable to ask why the Executive then ignored it.

Witnesses from other Air Stations confirmed they would fit a new stiffnut and bolt every time. The regulations require this when a weakened locking mechanism could lead to the loss of the aircraft.[41] It follows this applies where loss of life could result, but the aircraft survive. Here, the nut and bolt had been re-used. A perceived lack of resistance *might* explain why the technician over-tightened the nut. All Inquiries erroneously discussed the procedure for re-use, while omitting that these other witnesses had the correct information and training.

MoD then compounded its error. To test the continued effectiveness of a stiffnut when re-use is permitted or unavoidable, one must measure the remaining prevailing torque ('rundown torque'), against a minimum torque to be achieved. This is time-consuming, notoriously inaccurate, and requires special tools. And the smaller the nut, the greater the margin for error. RTI/Hawk/059 states: *'Spares and Special Tools - Nil'*. So, not only was the requirement for a new nut and bolt ignored, so too was the procedure for re-use.

The Service Inquiry noted new thread had been cut on the unthreaded shank of the bolt during assembly, evidenced by swarf (shavings of cut metal). The nut had been over-tightened to such a degree it had run out of thread. This is usually evident by feel, but the Inquiry did not comment on the failure to notice either this, or that the bolt had been bent. Nor, that this rendered both nut and bolt scrap.

To summarise, 100% replacement of the nut and bolt was mandated, yet

---

41 CAE 4000-MAP-01, Chapter 4.3.

the Instruction permitted re-use. In November 2011, post-accident, the mandate was restored. In November 2013, MoD decided to ignore it again. Maintainers must now be thoroughly confused.

Torque loading

The Service Inquiry criticised the lack of a torque setting in publications; repeated at the Inquest in 2014. (Logically, therefore, it should also have criticised the statement that no special tools were required to carry out the torque-loading). Most inferred failure by Martin-Baker. In fact, the publications were correct. A torque setting is inappropriate in this application, as the Drogue Shackle must not be compressed and must be free to swivel in the jaws of the Scissor Shackle.

The easiest way to think of this is... If one tightens the nut until it touches the lug, then neither compression nor pinching can occur. This was common practice among users, witness (e.g.) US Navy Technical Manual 55-1680-308-24, dated 1974:

'Tighten nut until contact is made with Drogue Shackle. Do Not Torque'.

Despite reminders after the accident not to torque load the nut (also reflected in Civil Aviation Authority Emergency Mandatory Permit Directives), MoD still insists on it being torqued to 50 lbf-in.[42]

This behaviour is key. Martin-Baker told MoD how to do the job properly. MoD didn't just omit a minor part of the procedure. It issued contradictory and manifestly dangerous instructions which contradicted good engineering practice. If a customer persistently ignores such advice and directives, and corporate knowledge is not retained, at what point does constantly repeating the information become futile? It was the Health and Safety Executive's case that, had the company (re)issued a warning between October 1990 and July 1992, then MoD would have heeded it and Sean's parachute would have opened. At any time, but especially given this evidence, that is one heck of a jump. In fact, it is arrant nonsense.

Fitting the nut, and 1.5 thread pitches

The nut was over-tightened. The Service Inquiry illustrated the resultant compression and out of parallel lugs, causing the shackles to

---

42 For example, AP 101B-4401-1G Maintenance Procedure 29-10/7.

pinch. (Figure 7). MoD claimed it was unaware of the need to avoid this. The Health and Safety Executive repeated this. In truth, MoD was not saying it *never* knew this; it was saying its policy was that it no longer *wanted* to know this.

Instead, the Service Inquiry claimed the nut and bolt were fitted in accordance with *'extant publications of the day'*, referring to 1.5 thread pitches of the bolt protruding above the nut. It cited this as good reason not to criticise maintenance practices. Yet, to achieve 1.5 pitches new thread was cut and the bolt bent.

While Defence Standard 00-970 indeed says *'at least 1.5'* thread pitches *in general use*, Air Publication 3279 'Standard Technical Notes' and the Naval Aircraft Maintenance Manual (both of which I was trained to) say *'at least one thread pitch'*. As does the US Federal Aviation Administration. Various RAF witnesses agreed.

But this was not *'general use'*. It was a special, safety critical application, with freedom of movement between the shackles required.

At the Inquest in 2014, a Martin-Baker engineer explained this during four hours of questioning. He reiterated that the 'special bolt' (so called because it, and its application, are non-standard) did not have the normal chamfer at the top of the thread, which ordinarily facilitates ease of fitting the nut. Here, the bolt was deliberately designed with no chamfer, but with a slightly domed end. This made starting the nut on the thread slightly fiddly, but the higher design aim was to avoid snagging the parachutes.

It will be apparent that on a chamfered bolt the end threads do not engage the nut locking mechanism (hence one reason for protruding threads), whereas on a non-chamfered bolt they do. Thus, the design deliberately and correctly eliminated the need to consider the *protruding threads* concept.

The witness also pointed out that it was normal to grind down protruding threads when using other nut types, for the same reasons. Invited to show this on a seat, he was stopped by the Coroner after pointing to around 60 Kaylock stiffnuts. The issue was dismissed there and then. Yet, as we shall see, the Prosecution later resurrected it, without being challenged.

However, while the above reveals differing MoD instructions, and a widespread lack of understanding, it is a red herring, erroneous in an assembly which must be free to not only swivel, but disengage. We always get back to this. Ultimately, the failure to ensure the shackles could disengage killed Sean.

*

No investigation mentioned what tools were used. Martin-Baker provide a tool kit for the Mk8/10 seats. It includes a tool commonly known as the 'flat spanner'; 3mm thick, to match and hold the smooth, thin head of the bolt. (The other end is used to release the Lower Harness Locks). There is also an open ended/ring combination spanner to undo/tighten the nut. When using the two spanners it is impossible to cut new thread or bend the bolt, because the necessary torque cannot be generated.

Post-accident, a party of MoD staff from the Aircraft Assisted Escape Systems team visited Martin-Baker. It was mentioned to them by the company that, if the correct tools had been used, the nut could not have been over-tightened. MoD replied that the spanners were not used - technicians used a socket and ratchet handle. Realising what it had just admitted, MoD clammed up and the matter was never raised again. Afterwards, both it and the Health and Safety Executive chose not to publish these two images, taken at the time...

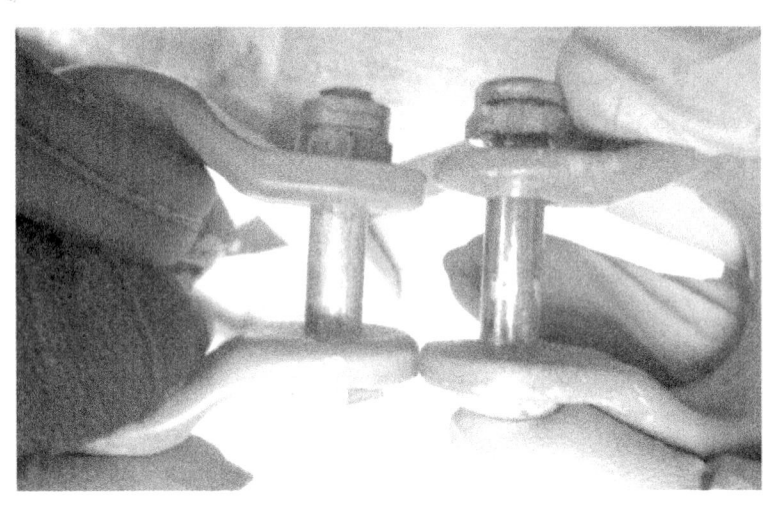

Figure 8 - On the left, XX177's shackle. The difference in the Inner Lug Gap is pronounced, and one can discern the bent bolt. *(Martin-Baker)*

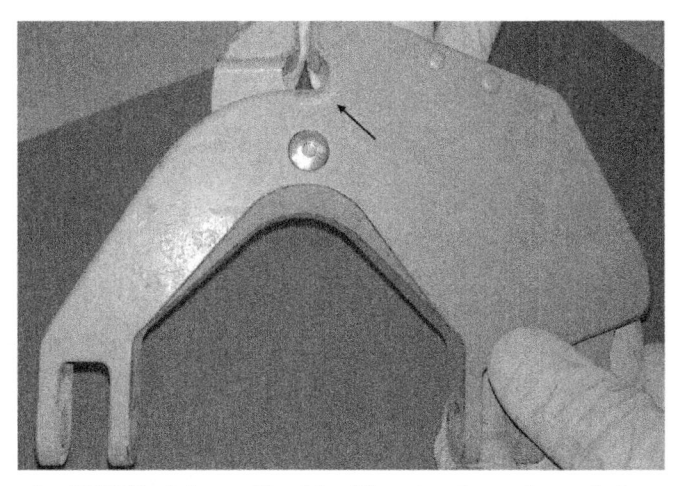

Figure 9 - XX177's Scissor Shackle. The circular rub mark (arrowed), caused by the over-tightened nut clamping the Drogue Shackle to the Scissor, was the first clue as to what happened. *(Martin-Baker)*

It can be seen Figure 8 depicts two different nuts. Martin-Baker specified a Cadmium plated thin steel 5/16" UNF/UNC all-metal stiffnut nut, to the A126-G66 Aircraft General Standard. This covers a number of types of stiffnut, but in this application only Philidas and Aerotight are permitted.[43] An Oddie type nut is strictly forbidden. As stated, XX177's shackle has a Philidas. The sample shackle has an Aerotight. The method of locking differs slightly, but the important point here is that one is thicker than the other. That nuts of different thickness *are* permitted demonstrates, again, there was *no design intent* to have threads protruding - entirely negating the Prosecution's argument.

Obvious questions arise. Why were the nut and bolt re-used? Was the tool kit available to the groundcrew at RAF Scampton? If not, on whose direction? If it was, who authorised the groundcrew not to use it? And why did supervisors allow it? Moreover, in March 2006 QinetiQ, in a Nimrod Fuel Leak Study Report, had noted out-of-date publications, *'inadequate tools'*, and *'aggressive use of metallic tools'*. Six months later Nimrod XV230 crashed, killing 14. As ever, MoD compartmentalised the issues, failing to ask if these serious failings were systemic.

---

43 AP109B-0131-5F, Servicing Note 5.

Figure 10 - Mk10B seat, illustrating the normal gap that exists between shackles. *(Martin-Baker)*

While MoD's report seems at first glance comprehensive, the omission of all this is alarming and plainly deliberate. The Panel made no attempt to check if the correct tools and techniques would have avoided over-tightening, and did not mention nut types. Instead, it concentrated on matters such as the nut being 0.007" longer than specification, without saying what the manufacturing tolerance was, or how tolerance build-up was addressed in the design. Nor did it mention that, while the special bolt was supplied by Martin-Baker, the RAF chose to select and procure the nut separately. Again, entirely exonerating Martin-Baker (because no single entity is now responsible for the Drogue Shackle Assembly). But the Panel's basic mistake was to assume the nut had to be torqued.

The RAF initially considered charges against the groundcrew, but withdrew. That might indicate the answers to the above questions were unpalatable, because of *who* would be implicated. For example, those who gambled on the dilution of training and support, knowing that the ALARP statement was now wrong.

Should the maintenance organisation have spotted the anomalies?

Charged with applying Routine Technical Instruction/Hawk/059, a tradesman should know that disturbing the system invalidated previous certification. Even if ordered to apply the Instruction, he should know the stiffnut and bolt must be replaced, not to torque load the nut, and that measuring protruding threads is not a valid serviceability test.

Before application, he is required to read and understand the Instruction. His training must already have ensured he understands the associated Air Publications, which warn:

> 'The instructions contained in this schedule do not absolve personnel from responsibility for acting upon circumstances which may come to their notice indicating the need for additional servicing'.

This placed a responsibility upon tradesmen (even if untrained on seats) to notify their supervisor that the Instruction did not ensure serviceability. To emphasise this, they <u>must</u> stop work immediately and report any anomaly.

The supervisor will not necessarily know the regulations forbid a Routine Technical Instruction in this application. But he will know it is at best incomplete, at worst dangerous. Precise procedures vary within MoD, but he will raise a Query Note or equivalent on his Engineering Department. They will contact the Issuing Authority, seeking a review of the data pack, and confirmation a Trial Installation was conducted to prove the Instruction. This would reveal that it was rogue. And so on, ultimately to the regulating authority, the Military Aviation Authority, whose role it is to prevent recurrence. But it had been actively involved in approval of the Instruction, so remained silent; compounding matters by abetting the Prosecution.

*

Crucial here is the demise of seat servicing bays. When the Hawk entered service, seats were removed and taken to the bay for inspection/servicing every six months. This was extended to 12 months, then 24 and 36. But, apparently, without considering the implications of Head Box servicing remaining at six months. One can readily identify when *in situ* dismantling and reassembly became the norm.

Even then, help was still at hand from the (few) remaining bay staff. But when the bays were closed in preference to centralised servicing, local expertise was swiftly lost. Deleting seat maintenance from armament technician training made the situation worse. Initially, if absolutely necessary, the RAF could cope reasonably safely with *in situ* servicing, as it retained the necessary corporate knowledge and skills to 'work around' the problem. But the removal of this defence meant the probability of the risk occurring increased. A slice in the Swiss Cheese had been removed.

Centralised servicing has pros and cons. In my experience there are too many cons. The main ones are loss of flexibility, loss of the mandated contingency factor (akin to scrapping War Reserves, which was also an RAF policy in the late 80s/early 90s), and the routine failure to assess impact. That is, it was often done for the wrong reasons, another *savings at the expense of safety*.

It is unimaginable the resultant risk was not raised by Martin-Baker, especially after their training contract was cancelled in 1983. The company's founder, Sir James Martin, who had passed away two years before, would have found this particularly galling. In the early days he had funded the training of Service personnel himself. His company mitigated the potential for error through a number of deliberate measures; including, by design, minimising disturbance once the shackles were assembled.

For its part, MoD had to ensure only fully trained and authorised personnel worked on the seat, training was consistent with maintenance procedures, and modifications and/or Instructions were properly developed, assessed and issued. MoD admitted it failed in all these duties. Worse, numerous declarations were made they *had* been fulfilled. MoD never discusses this, and legal authorities turn a blind eye. Aircrew relied on this false assurance, unaware that organisational failures and poor oversight had permitted these breaches of regulations, and the taking of shortcuts, to become the norm. Therefore, one must look further than the technicians.

Those familiar with the Mull of Kintyre case will appreciate the resonance. In 1993 RAF Odiham issued an illegal Servicing Instruction, authorising a dangerous act. (An in-flight check of safety critical fuel computer connectors). MoD has form here - Martin Baker do not.

10: Special information Leaflets (SIL) 704 & 704A

On 15 November 2011 Martin-Baker issued SIL704 'Drogue Shackle connection to Scissor Shackle maintenance check', emphasising it was not applicable to any seat fitted with a Gas Shackle. This subtly made the point that, had the Gas Shackle been fitted as recommended, Sean would almost certainly have survived. The check was to be completed *'before the next flight'*, and the SIL contained a prominent warning:

> *'During the operations detailed, <u>all</u> safety and maintenance notes detailed in <u>all</u> relevant technical publications are to be complied with'.*

That is, it was predicated on the seat being serviced properly. Also:

> *'Tighten the nut down until it contacts the Drogue Shackle ensuring no end float. [Axial clearance]. Make sure the thread is flush with or protruding from the end of the locknut, also ensuring the Drogue Shackle can be freely rotated about the Drogue Shackle bolt and with respect to the Scissor Shackle. To prevent possible binding on the Scissor Shackle, do not over-tighten or torque load the Drogue Shackle nut and bolt'.*

On 16 November 2011 MoD issued a corresponding Urgent Technical Instruction (Hawk/34), but it did not reflect the SIL. The most radical difference was an instruction to torque the nut to 50 lbf-in. Yet again it omitted to list the special tools required.

As it disagreed with Martin-Baker, MoD was required to notify its reasons to the company (and the Civil Aviation Authority). Where is the record of this process? What did Martin-Baker say, if notified?

Two years later, Martin-Baker replaced SIL704 with 704A. It added a warning that:

> *'Where there is a conflict between local instructions and the instructions defined in this SIL, this SIL takes precedence'.*

Clearly, Martin-Baker knew that MoD's *'local instructions'* did not reflect their own, and were warning of the dangers involved.

A corresponding Technical Instruction was issued on 5 December 2013. *'Before the next flight'* was replaced with *'earliest opportunity, but no later than seven days from receipt of the SIL'.*

Would Martin-Baker have been wiser to emphasise the SILs were raised in response to an accident caused by incorrect application of a rogue Routine Technical Instruction? Should they have been explicit? *We are not liable if we provide the proper instructions and training, but the customer (a) cancels the training contract, (b) ceases carrying out the instructions and/or instructs staff not to use them, or (c) refuses to implement mandated regulations. Nor can we underwrite the safety of a device that has been subject to a rogue MoD process.*

It is difficult for any Design Authority to adopt this tone in any Instruction, because it is letting everyone know that it doesn't trust its customer to do the right thing; which can make for fraught relations. But perhaps this case will lead to a rethink.

*

On 10 March 2014, Angus Robertson MP asked the Secretary of State:

'What steps it has taken to incorporate the revised checks into the appropriate maintenance publications'.

Philip Dunne MP, Minister for Defence Procurement, replied:

'Technical Instructions mandated checks of the ejection seat drogue shackle that were more stringent than those laid down in the SILs. In both cases, amendment action to the appropriate MOD maintenance publications commenced soon after issue of the Technical Instruction'.[44]

In fact, the new MoD instructions were manifestly unsafe, specifically the requirement to torque the nut. Concealing this meant it could avoid the question of who authorised this dilution of standards, and avoided the fact that the basic maintenance policy was flawed.

Mr Robertson also asked the Secretary of State to:

'Place in the [House of Commons] Library a list of Special Information Leaflets, with contents summary, issued between 1990 and 1996'.

Mr Dunne replied:

'Martin-Baker issues Special Information Leaflets (SILs) to the Ministry of Defence as a contract deliverable for each aircraft type. The Department does not therefore necessarily receive a copy of every SIL issued by the company. Information on SILs relating solely to aircraft types and equipment that are no longer in-service is <u>not held centrally</u> and could be provided only at disproportionate cost. A list of those SILs that are known to have been received by the Department between 1990 and 1996, with contents summaries, will be placed in the Library of the House'.

---

44 Hansard source (Citation: HC Deb, 10 March 2014, c32W).

MoD's own regulations require the audit trail to be held *centrally*, and the project teams maintain a SIL Register. Martin-Baker are contracted to, via the Post Design Services contract. The only variable is whether to seek maintained or unmaintained copies.[45] Unfortunately, the question and answer wrongly perpetuated the notion that a SIL was the only way of providing information.

To further illustrate how MoD sought to mislead Mr Robertson, on 25 February 2014 Mr Dunne had told him:

'The Mk10 ejection seat has been the subject of a mid-life upgrade programme, in the late 1990s, and an extensive modification programme from 2007 to 2010 that introduced many improvements. The Mk10 ejection seat in its current configuration is much improved as compared to 1996. As you are aware, further improvements are now being made as a result of the tragic accident to Flt Lt Sean Cunningham in November 2011'.

The author of this briefing would have known the reply only applied to Mk10A seats in Tornado - primarily, the fitting of the Gas Shackle.

*

These are not difficult engineering matters. They are akin to an aspiring pilot's first lecture on theory of flight, and the concept of lift. But the target audience has changed. 'Savings' have done away with those who knew how to manage the process. MoD placed itself in an impossible position. How could it convey to contractors that, because it no longer trained its staff properly, servicing instructions now had to be re-written so an untrained person could understand them? Not only that, but they had to describe prohibited acts. The concept is crazy. Nobody in MoD would be permitted to make such an admission. The necessary funding would (should) never be approved. And even if the requirement slipped through (or, increasingly likely, was signed by an unauthorised person), no reputable company would accept the contract. And so, the failures became endemic.

Martin-Baker may not have wanted to highlight the systemic nature of these violations, but others did. In 1996 the Director of Flight Safety warned that maintainers were being held *'hostage to fortune'* by inadequate information and poor training.[46] His words came home to roost - and not only on 8 November 2011.

---

45 Defence Standard 05-125/2, PDS Specification 20. (Form 500/778 procedures).
46 Puma Airworthiness Review Team report, paragraph 198, March 1996.

11: The Service Inquiry

The main difference between a Service Inquiry and Coroner's Inquest is the former seeks to determine the cause of the accident, the latter the manner or cause of death. To MoD, the death is an unfortunate outcome. Both aim to identify 'significant occurrences' with a view to preventing recurrence. Few achieve this aim.

Different approaches exist. The official position of the Chief Coroner in England and Wales is that, while there is a little overlap between Service Inquiry and Inquest, the Service Inquiry does not and cannot replace an Inquest. This reflects the law. However, in practice he allows individual Coroners to refuse to hold Inquests for deceased servicemen. Justice for military personnel and their families has become a postcode lottery.

In Scotland, the Crown Office and Procurator Fiscal Service defaults to the MoD view that a Service Inquiry negates the need for a Fatal Accident Inquiry (the Scottish equivalent of an Inquest). In this, it openly disagrees with Lord Cullen, former Lord President of the Court of Session, who chaired the inquiry into the Piper Alpha disaster. It has refused, for example, to hold an Inquiry into the deaths of three RAF airmen in July 2012 when their Tornado aircraft collided over the Moray Firth. Both jurisdictions are involved, because one of the deceased was repatriated to North West Wales. But MoD failed to notify the Coroner, meaning the mandatory Inquest was not convened. The Coroner claimed to be unaware, despite a very public military funeral. Even when advised by the public and the family, he refused to heed the mandate. (See also Chapter 29 for linkages).

This is not the place for deeper analysis of the Moray Firth case, but the clear aim was to avoid examination of MoD's false declarations to aircrew regarding the tolerable and ALARP status of mid-air collision risks. (In October 2021 the new Lord Advocate, Dorothy Bain QC, heard evidence on the matter from a former RAF engineer, James Jones. She replied that she saw nothing new that overturned the original decision. She was not influenced by other legal rulings that Service Inquiries are not independent, and that these rulings were *'not binding on the Crown'*; despite being made by legal Reviews set up by the Crown.

This situation is bewildering to legal experts and bereaved families alike.

*

The Convening Authority (in 2011 the Military Aviation Authority) must select the Panel immediately and set out its terms of reference, which are usually narrow. Normally, it will comprise a President, pilot and engineer, but they do not have to be experts in any particular field. The President is at least a Lieutenant Commander RN (Squadron Leader or Major), but must be at least one rank higher if there has been a fatality. Technical assistance is provided by the Military Air Accident Investigation Branch - in 2011 part of the Military Aviation Authority. While these officers may have direct experience of the aircraft type, almost by definition they cannot have the experience necessary to identify detailed or systemic airworthiness failings.

While evidence is given under oath, debate or cross-examination is forbidden. Any failure to tell the 'whole truth and nothing but the truth' is only revealed when a heavily redacted report is eventually released to the public, and can be analysed by independent experts. By then it is too late, the 'system' reluctant to regress. No Inquiry has been reconvened after serious omissions have been revealed. The most obvious candidate is the Chinook ZD576 case (Mull of Kintyre 1994), where the original Board of Inquiry and its findings were 'set aside' (quashed) in 2011, but not replaced. The desire to prevent recurrence is rendered hollow.

Service Inquiries no longer address culpability and possible negligence. Officially, this is a matter for the courts. But there is a need to apply military law, so the Inquiry may provide grounds for further investigation. This is not straightforward for the living, and unpalatable for the deceased. Often, in the absence of proof, sleeping dogs must be left to lie. This should not, however, prevent lessons being learned. But the line is drawn faintly. If an unlawful act is mentioned, and only one person was involved, then clearly that is where the finger is pointed. But reports don't say *this was an illegal act*. They merely discuss the final act, omitting that laws have been broken. As in this case.

From another angle, the report into the Sea King ASaC Mk7 mid-air collision (Iraq 2003, seven killed) contained criticism of civilian project managers. They were not interviewed, nor afforded the right of reply, even after producing written and photographic evidence proving the accusations wrong. The reports were not amended. Readers of the reports were deceived. And because Inquiries do not consider prior negligence, bereaved families from other accidents remained unaware the perpetrators were closely involved in other deaths, but protected.[47]

---

47 'Breaking the Military Covenant' (David Hill, 2018)

Seldom is a Panel so incensed it rebels. A rare exception was the loss of Chinook ZA721 (Falkland Islands 1987, seven killed). The Board of Inquiry President, Wing Commander (later Air Chief Marshall Sir) Malcolm Pledger, briefed aircrew he had been directed to find 'cause unknown', despite it being obvious from the physical evidence and confirmed by the Air Accidents Investigation Branch. This concealed Quality Control failures, protecting Boeing Helicopters. The evidence of this briefing was given under oath at the 1996 Mull of Kintyre Fatal Accident Inquiry held into the non-military deaths.[48]

*

One must differentiate between what the Panel reports, and what is released. They seldom align. In my experience Panel members are invariably excellent, but do not have the wherewithal or authority to ensure all evidence is disclosed. MoD is labyrinthine and few know who or what to ask. Also, a common criticism is that Service Inquiry reports are difficult to read, impenetrable to the layman and all but the most experienced in MoD. The mind-set is that no-one outside (in this case the Red Arrows and its immediate hierarchy) has any right to be reading it. This further militates against proper assessment, with the added 'benefit' of thoroughly confusing families. Non-disclosure agreements often make them unsure about seeking independent advice, serving to ensure MoD dictates the agenda at Inquests and other Inquiries.

In 1987, Mr William Tench, former head of the (then) Accidents Investigation Branch, prepared a report on behalf of the Government - 'Aircraft Accident Investigation Procedures in the Armed Services'. He was critical of MoD's procedures - in a word, they were unfair. It was buried. However, the tide turned when the Freedom of Information Act came into effect on 1 January 2006, followed swiftly by MoD being caught lying in the Coroners' courts over the Hercules XV179 (2005) and Nimrod XV230 (2006) accidents. These, and Chinook ZD576, were the subject of public campaigns (by the same few people). Each case was 'won' by exposing the truth - none of the aircraft were airworthy, but false declarations made they were.

A few Coroners, notably Oxfordshire's Andrew Walker and Wiltshire's David Masters, railed against being misled. In response, Government tried to curtail their powers by having military Inquests held behind closed doors. While this attempted cover-up was for the most part

---

48 'Their Greatest Disgrace' (David Hill, 2016)

unsuccessful, MoD remains both petrified and contemptuous of what it sees as a civilian process. It knows this is where families have their one opportunity to speak, so directs considerable resources at ensuring they are misled. The intent is concealment of wrongdoing. Today, MoD's default position is to refuse Freedom of Information requests relating to accident reports, routinely lying about accessibility of information.

Once, I believed the Chief Inspector of the Air Accidents Investigation Branch should provide independent oversight by reviewing and countersigning each report. However, its behaviour *vis-à-vis* the Shoreham Air Display disaster calls into question its independence from the Military Aviation Authority, forcing me to rethink. (See Chapter 27). Easily resolved, but why am I having to discuss ways of circumventing officially condoned misconduct?

The XX177 Service Inquiry

Director General Military Aviation Authority signed the Convening Order on 12 September 2012, 10 months after the accident - the timing dictated by the Lincolnshire Police investigation, which the Military Aviation Authority had been advising. (Bizarrely, and intriguingly, the police now distance themselves entirely from the case). The Panel was plainly disgusted at MoD's serial offences, dishing out a good kicking - albeit sometimes with the wrong foot.

Upon reading the report, I was struck by two things. First, why reassemble a safety critical escape system *in situ*, when serviceability cannot be verified? And, surely, someone realised the Routine Technical Instruction was rogue? This cast doubt on the extent of Martin-Baker's liability (if any), and the depth of the investigation.

Perhaps the best way to illustrate my point is that the Panel was required to assess and report on:

- Root Causes: event(s) leading directly to the accident which, if removed, would have prevented it. All were MoD liabilities.
- Contributory Factors: making the accident more likely. All 17 were MoD liabilities.
- Aggravating Factors: making the outcome worse. Of three, two were MoD liabilities, the other erroneous given MoD had the information it denied having.
- Other Factors: which are none of the above, but noteworthy in that

they may cause or contribute to future accidents. Of six, five were MoD liabilities, the other erroneous given the need to check shackle disengagement.

- Observations: issues irrelevant to the accident but worthy of consideration to promote better practices. All 13 were MoD liabilities.

Note: The Panel identified four Aggravating Factors, but the report lists only three. Similarly, 13 Observations but lists 11. One must assume the others were removed during review.

Of the report's 59 recommendations, 55 were already mandated. None applied to Martin-Baker. Recommendations any self-respecting engineer would understand include ensuring:

- Risks are tolerable and As Low As Reasonably Practicable.
- There is a robust and auditable method of tracking, reviewing and managing airworthiness decisions.
- Safety Cases are maintained.
- The Aircraft Document Set accurately reflects maintenance procedures.

Senior staff had declared all these activities had been carried out.

*

Most recommendations relate to maintaining the Build Standard, but the Technical Agency is not mentioned. Most roads lead to him, few lead away. These are (or were) one of the most carefully considered appointments in MoD, responsible for controlling and prioritising practical airworthiness tasks. By 2010 few in MoD could have the necessary background. Yet, such is their authority and responsibility, a Build Standard and Safety Case statement by the Technical Agency should be sought by every Service Inquiry. The reason is, MoD looks at three aspects - legal (and hence illegal), technical and airmanship. The Technical Agency has significant direct input to the first two, and to the training that facilitates the third. Nevertheless, the Panel, and those in MoD it did mention, should have understood serviceability issues, making the failure to recognise the breakdown of risk management so difficult to understand.

In 1992, MoD stopped maintaining the only procedural Defence Standard (05-125/2) setting out a Technical Agency's duties. While it remained mandated, the RAF ensured it would not be used by dismantling the specialist department who wrote and implemented it, and ridding itself of most postholders. It was withdrawn in March 2015 without replacement. Applying it would have removed the main root causes. It is structured around the 17 core components of the Build Standard. Some are clearly of more immediate importance than others, but if even one is compromised that is a danger sign. Here, at least nine had irretrievably broken down. The Panel identified the individual failures, but did not appreciate they were controlled by a single process, or that maintaining their integrity is a prerequisite to a valid Safety Case. No-one said, *Ah, one person is responsible for managing this and he is named in the contract, let's speak to him.* Perhaps the worst aspect is that all this had recently been explained to the Nimrod and Mull of Kintyre Reviews, and notified to the Military Aviation Authority in the presence of two Ministers.

*

The most serious violation noted is the lack of a Safety Case Report. The most serious omission is an explanation why it did not exist, and this not being identified as a factor of any kind. This is so obvious, I suspect the Panel's remarks have been removed because of *who* made the declaration it existed, and that they had personally checked it.

The Armed Forces Act 2006 sets out the offences. Section 10 (failing to perform duty), Section 18 (making false records), Section 36 (inaccurate certification relating to an aircraft or specified equipment). Each carries a maximum two-year sentence. By making the report public, was the MAA agreeing these offences took place? Or did it simply not appreciate offences *had* been committed? Either way, what does that say about the professional standards on display? Or the iniquity of pursuing Martin-Baker?

Admittedly, this is a difficult area for MoD staff. For much of my career the technical decisions relevant to this case would have been made by civilian engineers. Until around 1994, technical staff could not be promoted without satisfying a panel of one's ability to do *any* job at the next grade. (A test not imposed on non-technical staff or direct entrants). Overnight this became - *in time, and with training, one could perform reasonably well in the single post applied for.* An increasingly unbalanced two-tier engineering workforce was created, with experience cast aside. Who would impart the necessary training? A few dinosaurs, if MoD could bring itself to admit they were needed. It didn't.

Little wonder the Services sought to 'militarise' key engineering posts;

but that assumes they have suitably qualified and experienced staff waiting in the wings, which they don't. And one must understand which are key, this case illustrating all too well that they didn't. It seems no-one sought advice from anyone versed in practical airworthiness management. Partly because of the Service ethos to exclude civilians. But also because the Services are thoroughly perplexed by the vast variations in perceived experience and competence of their civilian colleagues at any given grade - something they are much better at managing.

*

The report reveals inexcusable failure of duty, describing regulatory and assurance systems in free-fall. For example:

> *'Commandant Central Flying School* [a Group Captain, and Delivery Duty Holder for the Red Arrows] *and his staff regularly flew with the Red Arrows and were aware that they did not carry out their checks in accordance with (regulations) and appeared content with this approach. The Commandant stated that "over time I think they have developed a set of routines which are absolutely fine from where I sit", indicating a normative acceptance of different standards for the Red Arrows compared to other military flying units'.*

A damning indictment. What is the point of a Duty Holder construct if the Duty Holder condones safety regulations being ignored? And even if his superiors were foolish enough to agree, all knew to revalidate Safety Cases against this new concept of use. So appalling is this failure, I think it the main reason why MoD admitted liability and settled with the family before the report was released. Had they seen it beforehand, the question of gross negligence manslaughter (against individuals) or corporate manslaughter (against MoD) would have arisen.

Moreover, important questions are left hanging. Were these unsafe practices brought to the Red Arrows, or did they develop there? And are they brought back to parent squadrons and perpetuated?

*

Perhaps the Panel's biggest mistake was:

> *'The assembly of the drogue and scissor shackle was found to be in accordance with...general engineering principles, and therefore did not contribute to the accident'.*

I need only note one basic breach of general engineering principles - there was no examination for the effects of friction.

However, the Panel rightly took aim higher up:

'The severity and frequency of signatory and supervisory malpractice on the (Red Arrows) was indicative of training and assurance shortfalls, compounded by a lack of supervision'.

But not high enough. Throughout, I quote direct warnings to a succession of very senior officers, including the RAF Chief Engineer at the time of the alleged offence. True culpability becomes clearer.

*

But one must be careful when discussing even the 1990-2011 timeframe. In that period, MoD's procurement and support arms underwent at least six major reorganisations. April 1999 was the watershed, introducing permanent, stove-piped Integrated Project Teams. Recognising the dangers, such teams had hitherto only been formed on a temporary basis. The first was established in 1989 to improve availability, reliability and maintainability of fire control and surveillance radars. The team leader (the project manager, who was also the Technical Agency) selected his team, who were seconded from their day jobs. He had complete control over collective funding.

In 1999 the new team leaders were more constrained and had less authority - yet were three grades/ranks higher. Most found it odd that some junior members of their team had previously been leaders of far larger teams; yet some senior members had never, for example, held any form of delegated technical or financial authority, a supervisory post, or even managed a project. The predicted effects of the new personnel policy had come to pass.

In 1999, staff were reminded that airworthiness regulations must <u>not</u> be applied. Much had changed in ten years and it was the first question I asked, reflecting our primary concern. What to do? Follow illegal orders, or meet legal obligations? Some chose wisely and were castigated. Others did not, and Sean was one of their victims.

A grey area?

The Service Inquiry's recommendations add weight to an 'initiative' proposed by businessman Bernard Gray. In his 2009 report 'Review of Acquisition', commissioned by the Labour government, he suggested transferring significant responsibilities to the private sector through a 'Government Owned, Contractor Operated' (GOCO) arrangement. In

2010 the new government appointed him Chief of Defence Materiel.

In a BBC interview on 27 December 2011, he mooted using *'Co-ordinating Authorities'* to implement this plan, omitting this was extant policy.[49] [50] When challenged by a Minister (Steve Webb MP) MoD claimed Mr Gray had said no such thing, distancing him from his own report.[51] This, when the broadcast was still available on the BBC's website. As proof of its wider utility, since 2001 the policy has underpinned the flagship Infantry programme (Integrated Soldier System), the primary aim being to reduce casualties. (Admittedly, no-one left in MoD actually realises this).

In April 2016, shortly after he left MoD, Gray was contracted to review whether the 'reforms' he had brought about had been effective. One might reasonably ask why MoD felt the need to pay someone to mark his own homework, when much of it was already mandated.

The Military Aviation Authority's conflict of interests

Vital to the probity of the investigation and subsequent legal proceedings is this, directed upon the Service Inquiry Panel:

> 'Should you consider that the decisions or activities of the Military Aviation Authority (MAA) could be a causal or contributory factor, you should report direct to PUS (Permanent Under-Secretary). In this event, you are to pause your actions and inform Director General MAA of your intention'.

The MAA was directly involved in a root cause, through attending Routine Technical Instruction/Hawk/059 development meetings. Part 1 (the Report) omits this, but Part 2 (the Record of Proceedings) mentions it; suggesting the Panel did not recognise what was a significant conflict of interests.[52] The Inquiry was not paused, and PUS was not engaged.[53]

The MAA's name is all over the decisions leading to the accident; exacerbated by one of its accident investigators acting as the Prosecution's main witness. The Health and Safety Executive and legal authorities must explain why they ignored this conflict. This collusion

---

49 BBC Radio 4 'Buying Defence', 27 December 2011.
50 Defence Standard 05-125/2, Post Design Services Specification 5, paragraph 4.9.
51 Letter D/Min(DEST)/PL MC00569/2012, 20 February 2012, from Peter Luff MP, Minister for Defence Equipment, Support and Technology to Steve Webb MP.
52 Service Inquiry report, Part 2, Exhibit 107, Attachment 1, Question 3.
53 Letter FOI2018/14060, 14 November 2018.

is sufficient to warrant re-opening the Inquest and police investigation.

The Red Arrows' view

Commanding Officer, Squadron Leader Jim Turner:

> *'It was a tragic accident that could have happened to any fast jet unit. I feel we've been singled out for some unfair treatment by elements of the Service Inquiry and I am very keen to prove to people we are as professional, if not the most professional, unit in the RAF. The treatment hurt me very much. It hurt all of us'.*

Clearly, he felt strongly enough to speak out. At their level this is what his men wanted - support from the 'boss'. However, it is prudent to point out his target was a report issued by his regulatory authority, whose job it was to review the Panel's report and remove any unnecessary conjecture. It was they - very senior officers - he was complaining about. Regarding *'elements of the Service Inquiry'*, again it is unclear if this was aimed at the same officers who had condoned the failings the report set out. Probably, like most, he was unaware of the notifications from RAF Directors of Flight Safety to the RAF Chief Engineers and the Air Staff throughout the 1990s.

Known risks

The grounding of the Hawk T.1 fleet was lifted on 8 December 2011. An RAF spokesman stated:

> *'The RAF chain of command have reviewed the safety evidence relating to the Hawk T.1 ejection seat and have no remaining safety concerns'.*[54]

Predictably, this proved untrue. In August 2018 I sought Part 2 of the Service Inquiry report under Freedom of Information. MoD declined, citing the time it would take (more than 3.5 days). MoD holds it as a single electronic folder and it would already have been redacted before being given to the family. I refined my request to 11 extracts, based on their description in the Exhibit List. Four were withheld and the others heavily redacted. As this would have taken longer than the original request, plainly there is something in the remaining papers MoD does not want to reveal.

---

54 https://www.flightglobal.com/news/articles/raf-clears-hawk-T.1s-for-flight-resumption-365739/

Included was the 22 Group Air Safety Risk Register summary. It is undated, but the last review was in October 2012. The next was set for April 2013. It listed three risks directly related to the final acts:

- Maintenance error due to human factors, raised 14 October 2011.
- Incorrect strapping-in leading to accidental ejection, raised 2 May 2012.
- Emergency equipment failure due to ejection seat being improperly installed, also raised 2 May 2012.

Post-mitigation, the probability of occurrence for the first is deemed 'remote'. The other two are 'improbable'. The terms are defined:[55]

- Remote - Likely to occur one or more times in 10 years, but not more than once a year.
- Improbable - Unlikely to occur in 10 years.

All three risks are deemed tolerable and ALARP. This implies that the (systemic) training, manpower and corporate knowledge failures noted by the Service Inquiry had been mitigated. Yet they remained, confirmed at the Inquest (2014) and in the Crown Court (2018).

We are asked to believe only one maintenance error due to human factors occurred across 22 Group in 2011. Its website claims it operates over 420 aircraft.[56] The figure was higher in 2011 as it included Air Cadet gliders (powered and unpowered), now largely unused due to maintenance errors and failure to maintain airworthiness. Could it be that the statistics now exclude any errors by contractors? And why was a standing risk only raised four weeks before the accident?

More straight forward is the assessment of incorrect strapping-in. Red Arrows pilot Flight Lieutenant Kirsty Stewart gave evidence at the Inquest saying she had committed the same error in 2003. She did not say what the RAF did about it when she reported it. Perhaps, given the wording of the risk, nothing, as she wasn't inadvertently ejected.

But the most obvious *'remaining safety concerns'* should be MoD's continuing support of staff who think it acceptable to have no valid Safety Case, and the regulatory authority's ambivalence.

---

55 Regulatory Instruction MAA RI/02/11 DG, Annex D, paragraph 3.
56 https://www.raf.mod.uk/our-organisation/groups/no-22-group/

Report release

Officially, Service Inquiry reports must be published *'as soon as possible after internal distribution'*.[57] On 13 April 2013, one month <u>before</u> it was published internally, the Crown Prosecution Service decided not to proceed against MoD. Part 1 was released on 6 February 2014, one week after the Inquest. Part 2, primarily detailed evidence and exhibits, was withheld. With all legal findings relevant to MoD issued before the report could be analysed, how comprehensive was any police investigation, and what did the Crown Prosecution Service base its decision on?

Upon querying this, Steve Webb MP received two replies from Ministers. On 17 March 2014 Philip Dunne MP stated:

'It has for several years been Departmental policy to publish Service Inquiry reports into fatal accidents at the conclusion of the Coroner's Inquest'.

And two days later Anna Soubry MP:

'The publication of Service Inquiry reports is considered on a case by case basis'.

In the first instance, the determining factor is whether or not a death has occurred. If it has, the report is not released until after the Inquest. If it has not, then it <u>may</u> be released subject to review by an unspecified process. The clear aim is to prevent independent scrutiny prior to the Inquest, thus allowing MoD to determine its direction. Precisely what happened in this case. If one simply asks for 'the report', only Part 1 will be provided. However, application is variable. Occasionally, a report will be released before the Inquest, usually in cases where there is no disagreement.

We will never know what conclusions the Panel came to. Its initial report was heavily diluted, evident by inconsistencies caused by failing to make complementary changes elsewhere. One can only imagine what was removed. I would like to think words to the effect *Not again!* Even so, the final report is a brutal exercise in self-flagellation.

As the Military Aviation Authority is aware of all that I have discussed, and had been reminded of it as recently as January 2011, one is entitled to ask why these lessons, repeated *ad nauseam*, are never learnt.[58]

---

57 Joint Service Publication 832 (Guide to Service Inquiries).
58 Meeting with Minister for the Armed Forces (Nick Harvey MP), 17 January 2011.

12: The alleged lack of technical information

The allegation against Martin-Baker

MoD stated it *could not find* the following warning, provided to five other customers *'circa 1991'*.

'To prevent possible pinching of the scissor shackle, which may cause hang-up of the drogue shackle during ejection, do not over-tighten or torque load the drogue shackle nut and bolt'.

It was claimed the warning was applicable when applying Routine Technical Instruction/Hawk/059. In fact, the intended applicability was during assembly of the Duplex Drogue. As the RAF had deliberately bypassed this intended application, then even if MoD did not have a warning (and, as you will learn, it did), this was unrelated to the alleged offence.

At the Inquest, and without explanation, this became *did not provide* - but no proof was sought or produced. The Health and Safety Executive later went further, claiming the information was not provided in <u>any form</u> between introduction of the Scissor/Drogue release mechanism (1952) and the accident. It knew this to be untrue.[59]

However, and to its credit, the Hawk Type Airworthiness Authority did conduct a search. Finding nothing in its own records it was thought, reasonably, that past records may have been archived to create space. But a thorough search of the RAF Innsworth archives by an RAF engineering officer and his staff proved fruitless. Astonishingly, not a single reference to Hawk could be found.

Casting his net further afield, it was suggested there may be some files in an old lean-to shed behind the Search and Rescue hangar at RAF Valley. There, dozens of boxes were found, but many were water-logged and mouldy as the building was derelict. A further two weeks were spent searching what remained, but the warning from Martin-Baker could not be found. But that is not the same as *never provided*, if only because many records were, in the officer's words, *'goo'*. In any other sphere of aviation, failure to keep repair records would result in revocation of certification.

(Notably, at paragraph 10A.118 the Nimrod Review recorded the same thing happening at British Aerospace, with much of the Nimrod *'deep*

---

[59] E-mail to Health and Safety Executive, 28 September 2016 15:49.

archive' held in a room that was flooded on more than one occasion. The XX177 Board did not comment on the status of British Aerospace's copies of the equivalent Hawk papers. Nor did the Convening Authority, the Military Aviation Authority (MAA), note that it was fully aware this failing was systemic).

This major breach of airworthiness regulations, requiring retention and maintenance of such records, forced the MAA to change tack and take heed of policy mandates. Having previously condoned these failings, it directed that correct control be re-established. However, and to be fair, in many parts of MoD similar airworthiness records were totally destroyed in 1991; including the files of the HQ Modifications Committees, who would be one of the first to see such a document. And when MoD(PE) relocated to Bristol in 1996, over 90% of project files were either destroyed or haphazardly archived. In other words, there could be no reasonable expectation of MoD ever finding the warning. Lacking such background knowledge MoD could not explain this, even if it wanted to. (And doing so would incriminate itself).

Later in this chapter I explain the process by which the warning would enter MoD, be promulgated and distributed. It will be seen there are many likely areas where the process may have broken down, all within MoD. I also explain that it would be unnecessary to provide the warning in the precise format mooted by the Health and Safety Executive, and that it may have been rejected by MoD on the simple grounds it already had the information. I set out why this makes the allegation unsafe.

Air Technical Publications - policy and practice

In MoD this is a centralised function. A single policy change affects all aircraft and equipment. With the advent of Aviation Publication 70 (AvP70) in 1970, MoD introduced a 16-Topic system. Nimrod MR1 was first to use it. Hitherto, the Air Ministry and other departments had a 6-Volume system, with aircrew publications (what became Topics 14, 15 and 16) managed separately. Hawk used the new system, but older publications (e.g. for common equipment) were not automatically converted. Contracts had to call up both standards. As the ejection seat design and its use straddle both periods, we must look at both systems.

Incorporation of the old Volume 4 data in the new Topic 1 caused most angst. Many had reservations. Mainly tradesmen, who understood that off-aircraft and on-aircraft instructions were now to be mixed, making it even more important to understand how equipment worked. One had to know that certain actions detailed in, ostensibly, an on-aircraft publication, were only applicable once the equipment was removed.

At the time, those planning the changeover could point to the superb training received by MoD's engineers, both Service and civilian. Trials of the new system showed they coped well. But that corporate and personal knowledge had to be maintained - a significant dependency not in the gift of Air Publications staff, who knew that spares, repairs, publications and training are the first target of any savings measures.

*

MoD has many 'Standards' for the provision of technical information. The Services, and even Units within the Services, have different requirements. There is no standard publications Standard.

Principal contractual and procedural requirements are set out in Defence Standard 05-123, which broadly describes publications procurement policies, nominates Publications Authorities for each of the standard range of Service publications, defines contractor liabilities, and indicates Standards to be used. The main one is the already mentioned AvP70 'Specification for Air Technical Publications'. Others include JSP(D)543 'Defence Technical Documentation', Military Standard 40051 'Preparation of Digital Technical Information for Multi-Output Presentation of Technical Manuals', and Aerospace and Defence (AECMA) S1000D 'International Specification for (Electronic) Technical Publications Utilising a Common Source Data Base'.

In 1978 the Army sought to streamline its procedures, yet its 'Aide Memoire to the Army Equipment Support Publications Management System' still listed over 20 related policies and specifications to be adhered to.[60] You get the picture - a minefield. A former chair of an AECMA International Steering Committee offered himself as a witness to explain all this, but did not even receive an acknowledgement.[61]

*

When the Mk10 seat was introduced in the late 1970s, Aircraft Technical Publications (ATP) was a major MoD(PE) department, employing over 500 Technical Authors at Professional & Technology Officer Grade III (PTO3, now Grade D). Sitting above them were Technical Editors. To

---

60 Aide Memoire 0100-P-002-013, July 1993.
61 E-mail to Clyde & Co. Solicitors, 20 November 2017 20:58.

place this in context, a PTO3 at a workshop would be a Divisional Manager in charge of a significant industrial and non-industrial workforce. This was generally one's fifth grade. Today, recruits skip these five grades, never learning what is not being done. Appreciating this dilution of standards is essential. It helps explain why MoD lost its understanding of what was perfectly good information.

At the time of the alleged offence (1990-92), ATP was transferred to Air Member Supply and Organisation, which became Air Member Logistics in 1994. Henceforth, and like all airworthiness related activities, their work was considered a waste of time and money. Annual cuts of ~25% were applied to the budget - an ethos that prompted the many warnings I have noted from the Director of Flight Safety, including this to the RAF Chief Engineer in 1996:

> *'There is a gap in the orders and procedures concerning the amendment of Air Publications. The problem lies with the question of what the tradesmen do in the meantime. Do they work to and sign for an activity which is known to be wrong, or do they work outside the content of the maintenance document and thus be hostage to fortune should a problem occur?'* [62]

This repeated notifications of poor communication between MoD departments, resulting in safety information not being disseminated - precisely what Martin-Baker were accused of. It supported long-standing complaints by civilian staff since 1988, endorsed by the Director of Internal Audit.[63]

It is worth setting out some of these prior warnings to the Chief Engineer. (Remember, these apply to all aircraft). From the Chinook Airworthiness Review Team report of 1992:

- *'There has been a lack of focus on certain critical issues, such as the inadequacy of publications'.*
- *'The Support Authority has been compelled to defer general airworthiness improvements, such as a comprehensive update of publications'.*
- *'Attempts to alleviate the problem have been hindered by inadequate resourcing over a protracted period'.*
- *'RAF tradesman have been working with poor publications for far too long'.*

And the Tornado Airworthiness Review Team report of 1995:

---

62 Puma Airworthiness Review Team report, paragraphs 197 & 198, March 1996.
63 D/DIA/5/295/10, 27 June 1996 Director Internal Audit report 'Requirement Scrutiny'.

'These conflicts point to a clear need to conduct a detailed review of all documents associated with flying against the one authoritative document, the Release to Service. The size of this task should not be underestimated'.

The Nimrod Review of 2009 characterised this as the *'heyday of RAF Logistics Command'* and the *'golden period'* of airworthiness.[64] It praised two of the three Chief Engineers from this time (1991-98), despite the Chief of Defence Procurement, Sir Robert Walmsley, confirming to the Public Accounts Committee on 3 March 1999 that the problems remained.

*

Historically, companies would prepare servicing manuals. RAF personnel at the Air Ministry Servicing Development Unit - from 1953 the Central Servicing Development Establishment (CSDE) - would adapt them for Service use. There was nothing wrong with the originals, they were just aimed at a different audience. Companies initially prepared them for their own highly trained staff. Martin-Baker's specialise in ejection seats and little else. Unlike servicemen, they are not diverted for years on end to another unrelated domain, struggling to recall training when they return, and having to learn new procedures and even new technologies. Also, one must consider operational aspects. What if, in theatre, a squadron's armourer is incapacitated? Someone untrained in seats, but with trade training, must do the work. He must have a publication spelling out each step; but would (or should) not, for example, need to be told how to fit a stiffnut. It can be seen how crucial it is that MoD informs industry as to the precise training military personnel have; and that contracts be amended if this changes.

There are various ways of managing this. Generically, MoD sets out its requirement thus:

'The depth of descriptive information shall be dictated by skill level and supported by both unit and overall system diagrams such that it meets the requirements of the operator and Engineering Maintenance Policy as defined by the approved Logistic Support Authority.

Where shown to be cost-effective, existing Commercial-off-the-Shelf publications, which should be to a good commercial standard, containing the necessary depth and scope of information, may be acceptable instead of the

---

64 Nimrod Review, paragraph 13.124, 28 October 2009.

stated requirement, with the exception of Topics 14, 15 and 16'.[65]

Ultimately, MoD states to industry: *This is what we will do or provide (Support Policy and level of training), and you must meet us at the 'control boundary' to avoid gaps.* This is called Interface Definition, and is the basis of all systems integration and functional safety. In this case the 'system' is Publications Management, part of the Safety Management System. Failure to adhere to this mandated principle is thus a systemic failure. If this changes, as it did with the introduction of centralised servicing and reduction in training, then compensatory provision must be made to update the publications. This was effectively prohibited due to the aforementioned annual cuts.

In 1995 the RAF Director of Flight Safety reported:

'Evidence from both MoD(PE) and Support Authorities indicates there is an airworthiness hazard in continuing to make staff reductions without full consideration of the task. Resources must meet current workloads'.[66]

The cuts intensified, followed in 1997 by the Chief of Defence Procurement announcing over 500 specialist engineering posts were to be chopped at the new procurement HQ in Bristol. No investigation sought witnesses who recalled this era. When a former senior official in Air Technical Publications (and before that an RAF engineer and university lecturer in aircraft instrument systems) came forward, the Health and Safety Executive dismissed his evidence without hearing it.

The person who controls this at the detailed level is the Publications Authority, invariably a civilian in Aircraft Technical Publications. Unsurprisingly, his requirement is recorded in Interface Definition Documents. The process is made much easier, and safer, if the contractor (Martin-Baker) is awarded the contract to train MoD staff. MoD cancelled it in 1983.

At a policy level this is overseen by the Provisioning Authority, a Service HQ post. The last disappeared in 1988 and it is unclear today who is responsible; and equally clear the work is seldom carried out.

*

Gradually, CSDE's work was contractorised and MoD workshops privatised. The recruitment ground for more senior posts requiring airworthiness delegation was gone. The criteria for 'suitably qualified

---

[65] DDAL(RAF) Air Technical Publications Statement of Requirement, 30 November 1995.
[66] Tornado Airworthiness Review Team report, paragraph 180, November 1995.

and experienced' was so diluted as to be unrecognisable (and remains so) but contracts with industry were not amended to compensate. By 1993 Aircraft Technical Publications was decimated and merely a small Deputy Directorate within the RAF. Only at first did it retain an Armaments Section, manned by former armourers. CSDE, too, remained in a smaller form, continuing to write Maintenance Schedules. When industry prepared them, CSDE still did the technical vetting and were seen to approve them. That is, they checked technical accuracy, not style or formatting.

That the parachute release mechanism was introduced in 1952, and Red Arrows groundcrew were working on an aircraft introduced in 1979, meant the possibility of <u>contradictory</u>, not missing, instructions was always going to be a factor in this case. The rogue Routine Technical Instruction proves the point. But neither the Service Inquiry nor Health and Safety Executive went there.

*

At various points in a project MoD formally states it is content with the information provided. Most come under the general heading of Configuration Reviews, during development. In practice, detailed vetting of publications is carried out on a rolling basis by MoD's Publications Authority. Ultimately, the contractor supplies certified data, to be used during trials; and its use is verified during servicing. To facilitate this, the company is required to provide equipment, assistance and facilities. (What precisely it provides is determined by Intellectual Property Rights ownership, something overseen by the aforesaid, and now non-existent, Provisioning Authorities).

<u>The</u> crucial milestone is Transfer to Post Design Services, whereby the design is brought Under Ministry Control. That does not mean MoD takes over design responsibilities. Rather, it is the point at which the designer can no longer unilaterally change a design. Put another way, MoD now chairs the Configuration Control Boards. It must provide a chairperson (the Technical Agency) who is expert in all aspects of the equipment's manufacture, support and use. A basic component is tabling all associated publications, and a formal statement by the Service that it is content they are comprehensive, accurate, and fit for purpose. Thus, the contractual and Build Standard baseline is set, and the above 'control boundary' verified. The documentation suite is then maintained throughout the life of the equipment, in accordance with the mandated Defence Standard 05-125/2. As MoD pays for the work, it

dictates what is done. I have mentioned this Standard often, and there is a reason. If it is suspected that any of the above process broke down, to stabilise the process one 'simply' implements the Standard. This applies universally, not just to publications.

All this was ignored by the Health and Safety Executive, its entire case based on an assumption that Martin-Baker had sole responsibility. In fact, once the design was Under Ministry Control, their contract restricted them to submitting proposals, including warnings, which MoD could accept or reject. Here, we know that the RAF rejected its own Maintenance Policy - a poor starting point for a prosecution of anyone but MoD. In 1984 it then rejected Martin-Baker's proposal to eliminate the risk of shackles failing to disengage. It can be seen why a warning reminding MoD of the risk would, at a certain level, be unwelcome in 1990-92. And certainly not circulated, because it would reveal the old seat design was no longer considered ALARP by the designer.

*

Companies do not supply MoD units with copies of documents. They are contracted to supply Camera Ready Copy (today, Print Ready Copy); for example, in accordance with PDS Specification 19. Under a separate contract, often with Her Majesty's Stationery Office, it is printed in the required quantity and distributed by MoD. During the period in question, on 12 May 1991 MoD Telford relinquished responsibility for this, handing over to MoD Central Services Establishment (CSE) at Llangennech. Again, both periods must be assessed to see the whole picture.

In 1995 the RAF Director of Flight Safety noted CSE was under-resourced. This affected all aircraft. Also, that CSE's official policy was to tell Air Stations to photocopy publications and distribute copies themselves. He reported:

> 'The provision of Air Publications appears to be haphazard. CSE have little understanding of the significance of their task...and must employ working practices that provide an audit trail'.[67]

This further example of problems caused by constant upheavals and cuts emphasises how difficult it was for investigators to establish who did what. What is certain is the company's role was confined to the early part of the process. The courts were misled into believing only Martin-

---

67 Tornado Airworthiness Review Team report, paragraphs 151 and 152, November 1995.

Baker could be responsible for 'missing' information. On the contrary, the Director of Flight Safety's words, condemning official policy, render any accusation against the company unsafe.

Difference Data

This subject is central to the entire case. Examining it reveals the process by which Martin-Baker provided the necessary information.

MoD policy is to keep information sources singular, meaning (as far as possible) only one document need be amended if data changes. Once information is provided, only details of differences are required. A Difference Data Manual provides information on variations between configurations. Only when a particular variant of an item leaves Service is unique data removed from publications.

The parachute release mechanism in the Mk2 seat was different to Mk1, introducing the Drogue/Scissor Shackle arrangement and an improved Drogue Gun. In 1952 Difference Data was required to update schedules and training. This was incorporated in RN/RAF training and Air Publications.[68] When work commenced on modifying the existing seats in August 1953, RN maintainers were trained by Martin-Baker and carried out the work at Air Stations. By definition, therefore, the Ministry was content that the information provided ensured serviceability.

The next design iteration, the Mk3 seat, introduced the Duplex Drogue, which incorporates the Drogue Shackle. This meant the Drogue Shackle, and its nut and bolt, were assembled in a different servicing bay. Similarly, updated and satisfactory information was provided and incorporated in publications.

And so on until 2007, when the Gas Shackle replaced the Scissor Shackle in the Mk10A seat. At that point, Difference Data was required for the Gas Shackle. But the 1952 data would be retained, as the Scissor/Drogue design remained in other seats (not only the Mk10B). However, as some other users (countries) had been using the Gas Shackle exclusively since 1984, between 1984 and 2007 *their* requirement was completely different to MoD's. Again, straddling the period in question.

This procedure mitigates against loss of corporate knowledge (but not experience), the main driver behind all technical authorship. Retention

---

68 For example, AP108B-0107-1, paragraph 16(7), at AL14, September 1991.

of that data is MoD's responsibility. If it loses it, it must ask for it again. It is not for the company to dictate what MoD does with it, or monitor MoD's compliance with its own regulations.

An investigator must carefully work back down this evidence trail. Here, culpability quickly becomes clear. But the Health and Safety Executive did not look, immediately blaming Martin-Baker. When taken to the evidence, it deemed all of the above *'irrelevant'*.

Origin

The Prosecution case centred around how Martin-Baker dealt with two different technical queries.

First, in January 1990 McDonnell Douglas questioned the torque figure when tightening the Drogue Shackle nut on a Mk10L (lightweight) seat in a US Navy F/A-18 Hornet, and the clearance between the Drogue Shackle lugs and Scissor Shackle. It was not stated what information source they were referring to, only that they were querying whether Martin-Baker agreed with the US Navy removing its self-imposed requirement to torque load the nut.

The F/A-18 entered service in 1983, the Mk10L having the standard Scissor/Drogue Shackle arrangement. By 1990 even the latest Mk10 design was considered dated, the Mk14 Naval Aircrew Common Escape System (NACES) seat having entered service the previous year with the US Navy. This featured a sensory system and 5-mode electronic sequencer to allow it to adjust for altitude and speed. The Drogue was in a separate container and operated by the sequencer, eliminating the old shackle arrangements, be they Scissor/Drogue or Gas. In 1990 McDonnell Douglas and US Navy would be used to the new seats, and may have lost some corporate knowledge. The introduction of electronic sequencing, and in newer NACES models event recording, brought its own problems - electronic component obsolescence - requiring frequent upgrades, which tends to divert resources from older designs. MoD was the same.

Perhaps confusing McDonnell Douglas was that US Navy technical manuals correctly said *Do Not Torque* - consistent with standard engineering practice and Martin-Baker's training.[69] Like MoD, the US Navy adapted Martin-Baker's manuals and applied their own standards.

---

69 For example, US Department of Defense Technical Manual TM 55-1680-308-24.

(Explained later). McDonnell Douglas may have been referring to a rogue US manual, or an isolated directive not to implement the manual. (Much as the RAF decided not to heed the same advice).

Martin-Baker drawing office employees were incredulous at the question, the answer to which was *Do what we've already said*. In 2018 the Judge, Mrs Justice Carr, confirmed:

> 'The risk correlates to the US Navy's requirement to torque load the locknut, which <u>did not originate from Martin-Baker or apply to the RAF</u>'.

That is, the query was irrelevant to this case, the Judge noting there had never been such a query before.

Given this situation, if Martin-Baker even considered advising MoD, what would they say? *One of our customers made an erroneous query about torque?* What would MoD say? *We already know what to do, it's in the training you gave us.* The problem was, those who would reply on behalf of MoD, armourers in Engineering Authorities, may not have worked on seats for some years and might be unaware of the scale of skill fade. Matters got even worse when, shortly thereafter, these staff disappeared, never to be replaced, and instructions were issued <u>not</u> to comply with the manuals or training. But that was MoD's problem to manage, not Martin-Baker's.

The key exculpatory fact is that the RAF instructed maintainers to torque to 50 lbf/in, reiterated after the accident.

<u>The Judge dismissed this component of the Prosecution's case</u>.

*

Second, in June 1991 British Aerospace queried how to assemble the Drogue Shackle nut and bolt. The issue was that of achieving *1.5 threads protruding*; which we have established is erroneous in this context. The Judge's remarks reveal that, again, there was confusion within Martin-Baker, because their design intended the bolt end be at least flush with the top of the stiffnut, with the nut just touching the Drogue Shackle.

British Aerospace were mistakenly thinking in terms of thread count - which persisted, as they still insisted it was *1.5 threads* during the investigation in 2011/12. The main concern in Martin-Baker would be that the company did not think in functional safety terms. A further implication being that seats serviced by British Aerospace under their maintenance contract may be exposing aircrew to risk.

An investigation commenced. There were two elements to consider -

the release mechanism design, and servicing instructions. In August 1991, an MoD Quality Assurance Representative at Martin-Baker, Mr Barry Cowell, signed off the report, confirming he was:

> *'Satisfied that the outcome to the British Aerospace inquiry had been achieved in accordance with the correct procedures'.*[70]

I've been in this situation. Often the query is vague, because the originator doesn't fully understand - which is why he asks. My colleagues and I would discuss queries and jot down notes as *aide memoires* - not as formal answers to the query, which had to be counter-signed by the Leading Draughtsman or Engineering Manager. Here, the Judge based her remarks on a Martin-Baker engineer's initial thoughts - his jottings on a compliments slip. His primary concern - more sheer disbelief - was that anyone would over-tighten a nut in this application.

In court, on 13 February 2018, Martin-Baker's QC made the point that the Health and Safety Executive did not grasp Mr Cowell's role and the significance of his signature. That is, he was challenging the value of an investigation that ignored inconvenient facts. He received no reply.

The thread count issue was not germane, and quickly resolved. Mr Cowell's written declaration rendered the query irrelevant to the allegation. And by naming him the Judge confirmed she knew the charge against Martin-Baker had no merit. But this time she did not dismiss it.

What information did MoD ask for?

An obvious question, unasked by any Inquiry. Each customer decides its own support policy. A user may have Martin-Baker do everything, so require little in-depth information. MoD established its own servicing bays, so required more detail and specialised training. Hence, and again, information provided to one customer may not be required by, or suitable for, any other. The Health and Safety Executive made much of five customers receiving a warning, but did not mention most did not. Nor at any point ask why not. The reason is - they either did not need the detail or, like MoD, already had it.

Moreover, up to and beyond the introduction of the Mk10 seat, different arrangements existed for maintenance information. For example, prior to 1990 Defence Standard 00-52 (General Requirements

---

70 Sentencing remarks of the Honourable Mrs Justice Carr, 23 February 2018, page 18.

for Product Acceptance and Maintenance Test Specifications and Test Schedules) specifically excluded maintenance documentation. On 26 January 1990 an Interim Issue 2 was circulated for comment, mooting for the first time a formal requirement for maintenance documentation. It was only on 27 September 1991 that this was issued, but not retrospectively applied to existing contracts. Yet again, different policies prevailed during the period in question.

What information did Martin-Baker provide?

The pivotal question, conspicuously avoided by MoD and the Health and Safety Executive. The earliest Martin-Baker manual available to me is M-B/17 'Mark 4 Series Ejection Seat - Instruction and Servicing Manual' (c.1956). The Mk4 was fitted to (e.g.) Hunters, Buccaneers and Lightnings. Like any proper servicing manual, it details what should be carried out, when, why and how. At paragraph 91 it states:

'It is necessary to remove the seat from the aircraft...'.

And paragraph 107b, covering later testing in the servicing bay:

'Observe that the mechanism has operated and has removed restraint from the Scissor, thereby freeing the Drogue Shackle'.

While the Mk4 seat differs from the Mk10 in that it is not of modular design, both have the same shackle arrangement. Therefore, the important principle remains that the final check for serviceability and functional safety is disengaging the shackles. Investigations should have examined and explained the process by which MoD transitioned from this disengagement check, to waiving it.

These instructions were reflected in Air Publications. The Bay Servicing Schedules of 1992 require functionality to be confirmed by *'manual testing'*, as above.[71] Again, this is sufficient to refute any suggestion Martin-Baker didn't provide adequate information. Not only that, but the same schedules are mentioned in the Air Accidents Investigation Branch report into the 2015 Shoreham Air Display accident - a report which both MoD and the Health and Safety Executive contributed to. Proving, again, that both were in possession of exculpatory evidence before the trial of Martin-Baker.[72]

---

71 AP109B-0131-5F.
72 Shoreham Air Display report https://www.gov.uk/aaib-reports/aircraft-accident-report-aar-1-2017-g-bxfi-22-august-2015

If the shackles do not disengage when the scissor opens, action must be taken to ensure they do - and the primary cause is pinching, caused by over-tightening. Only then may the seat be certified serviceable. Thereafter, the release mechanism must not be disturbed without repeating this check. As the design remains the same in the Mk10B, so too does this principle.

It can be seen that even if the warning was not available (which is not the same as *'not provided'*), that was not a root cause. The failure to check serviceability was. The risk was created when this mandate was waived. Again, the Health and Safety Executive deemed all this irrelevant, denying any commonality between the Mk4 and Mk10 seats.[73]

Who was Martin-Baker required to inform?

Another important question which the Health and Safety Executive did not ask.[74] MoD's simple and effective policy is to provide a single point of contact - the Technical Agency. To the company, he is MoD. He controls the design and appoints his opposite number at the company - the Post Design Services Officer (PDSO). This is a unique authority, reflecting the granting of financial delegation to the PDSO so that he may commit MoD funding to urgent safety tasks without first seeking approval. In this way, safety issues can be addressed within minutes - but only if these mandated procedures are implemented.[75]

The Technical Agency, like the project manager, must be able to carry out the roles of everyone in a project team, regardless of grade, rank or specialisation. (Very few of the latter can, explaining many of MoD's procurement problems). One may be Technical Agency and project manager. If separate appointments, once the design is Under Ministry Control the Technical Agency has primacy on design and safety. The important thing is, he is a named appointee with airworthiness delegation, so readily identifiable.

Few are resident at companies, so are often represented by a resident Quality Assurance Representative (QAR). (The Publications Authority is a QAR and Technical Agency in his own right). All three must have an understanding of the procedures, processes and technologies at the contractor, and be competent in review techniques. A resident QAR is

---

73 E-mail from Health and Safety Executive principal investigator, 17 May 2018 12:36.
74 E-mail from Health and Safety Executive, 7 February 2019 14:39.
75 Defence Standard 05-125/2, Chapter 1, paragraph 1.4 & PDS Specification 5.

placed in a position of enormous trust. Like the Technical Agency, he is required to base judgments on substantial objective evidence provided by the contractor, thus proving the effectiveness of its Quality System. Subjective evidence or opinion is not enough.

This obligation is the best indication of what Mr Cowell was signing for in 1991. (He was MoD's highly regarded expert on Aircrew Assisted Escape System explosives and propellants, having previously been in charge of a 100+ strong workshop - suggesting a full understanding of the issues). He was confirming he had seen, understood and agreed with a compelling body of evidence supporting the final response - <u>not notes on a compliments slip</u>. By definition, this included Martin-Baker's reasoning behind the Scissor/Drogue design, and its servicing instructions - because the design is incomplete, and cannot be accepted off-contract, or into service, without those instructions.

Moreover, MoD had previously signed the Certificate of Design saying it was satisfied with the information provided.[76] The Certificate includes a list of all tests conducted to show compliance with the specification. As the specification (obviously) requires the shackles to disengage, the Certificate cites the information necessary to ensure this. And, as the system incorporated explosives and propellants, additional MoD signatures were required. No deviation is allowed from this procedure without the express approval of the Treasury Solicitor. No evidence was offered in this regard.

Why was Mr Cowell signing a report arising from a British Aerospace query? <u>Because Martin-Baker had correctly notified MoD</u>. This is precisely the kind of empirical evidence required by investigators, and again indicative of the lack of depth of all investigations. It can be seen that he was, in fact, carrying out his obligation to ensure the Certificate of Design remained valid.

Martin-Baker were accused of not advising MoD of the issues addressed by this investigation. MoD signed the report. QED.

The decision point - what information enters MoD?

Having notified Mr Cowell of the issues, the matter was now outwith Martin-Baker's control. It was for MoD to decide whether existing information had to be amended, and how to disseminate it.

---

76 Controller Aircraft Instructions, Chapter 2.3.2.

What is the process by which the decision is made? The gateway is the Local Technical Committee, the formal name for Post Design Services meetings, and chaired by the Technical Agency. It allocates and prioritises tasks, and records progress and decisions - which by definition are all safety related. The final question is always - *Has the Build Standard been maintained?* This is why Mr Cowell made his formal declaration. It was linked directly to the Safety Argument and the Master Airworthiness Reference.

MoD attendees include the Provisioning Authority, the 'owner' of any given range of equipment. (For example, one person will 'own' all Radar or all Navigation equipment). He tends to receive most actions as anything involving policy, future funding, availability, reliability or maintainability falls to him. (In practice, he delegates attendance at most meetings to the Engineering Authority). It is he who prepares and staffs any case for resources to maintain capability, or compensate for its loss. In short, he is the Service's trouble-shooter on support matters. One can perhaps appreciate what is <u>not</u> done if there is no Provisioning Authority. Since 1988 the work has only carried out if someone has the gumption, but not necessarily the training, to do it. One inevitable effect can be seen in Routine Technical Instruction/Hawk/059. Today, 'Requirements Managers' and 'Integrated Logistic Support Managers' are the nearest equivalent, doing some minor elements of the job.

*

Mr Cowell's declaration reveals Martin-Baker and, at that point, MoD, acted correctly. What happened next is where investigations needed to look. The Technical Agency would have asked - *Does this change anything we already know?* It is entirely likely that, satisfied MoD already had the information, he decided 'no further action'. But, demonstrably, the information was provided to a gatekeeper. And it can been seen that the information need not necessarily be on a piece of paper headed 'Warning'. Providing it within a wider report is fully acceptable; but of course reliant upon the retention of sufficient MoD corporate knowledge.

If the Technical Agency agrees the information *is* needed, he tasks the company to deliver Camera Ready Copy after MoD vets and verifies the proposed content. But simply providing the information, <u>by whatever means</u>, is not enough. It must be 'design incorporated'. A period of administrative lag is permitted - measured in, at most, weeks. In the interim, advance notification may be issued by signal. However, in 1990 it became common practice to 'roll-up' amendments annually, due to the denial of resources. In 1995, the RAF Director of Flight Safety notified the Chief Engineer and Assistant Chief of the Air Staff that the enforced delay was now around two years.[77] These addressees reveal his primary concern - airworthiness.

What was MoD required to provide to Martin-Baker?

Yet another unasked question. One procedure was set out in Defence Standard 05-125/2. (By now you'll perhaps appreciate this is the bible. Above all else, applying it is what would have saved Sean). MoD is required to provide Certificates of Correctness of Engineering Data; for example, as supporting evidence for the aforementioned Transfer to PDS. Part A is a certification to the contractor that:

'MoD has checked the information contained in the designated lists, drawings, documents, etc. is correct and that they are complete'.[78]

This applied to:

'Equipment which has been accepted by the departments of the Controller Aircraft and the Master General Ordnance'.[79]

The contractor then certified to the Technical Agency, in Part B, that the information was a true record. The Certificate was supplied with each Master Design Index for a complete equipment, assembly or sub-assembly. By issuing the Hawk Release to Service, the Assistant Chief of the Air Staff made a binding declaration this audit trail was complete.

Asked to demonstrate accuracy of information forming the basis of the Safety Case, one produced *(inter alia)* the Certificate. As the allegation against Martin-Baker centred on what information MoD had, the Certificate should have been one of the first pieces of evidence requested. In late 2018, I submitted a Freedom of Information requesting it. MoD claimed not to recognise the term:

'For Air Systems, the current standard of regulatory documents are Regulatory Articles produced by the Military Aviation Authority. Defence Standard 05-125/2 is not listed'.[80]

Typical of MoD replies, this is irrelevant to the period in question. These

---

77 Tornado Airworthiness Review Team report, paragraph 150, November 1995.
78 MoD Form D1/PDS 'Certificate of Correctness of Engineering Data'.
79 Defence Standard 05-125/2, Preface, paragraph 1.
80 Letter FOI2018/15965 from DE&S Secretariat, 18 January 2019.

Regulatory Articles, introduced from 2010-on, do not mention this requirement. Yet again, a basic airworthiness procedure has been abandoned without considering the consequences. Once more, MoD's inability to produce a key part of the audit trail renders any accusation against Martin-Baker unsafe.

The link between training and the accident

As the design arrangement and function of the release mechanism remained the same from Mk2 to Mk10B, the original data remained valid. This was imparted to tradesmen and aircrew via training contracts with Martin-Baker. Back at their squadrons and workshops they used this knowledge in conjunction with CSDE's publications. The two, training and publications, go hand in hand.

Training is a risk control measure. When the contract was cancelled in 1983, the Safety Analyses needed to be revisited to examine the effects of the change. What did MoD do? The Service Inquiry provides the answer. More posts were cut and training curtailed. There was no training course to qualify Red Arrows groundcrew to work on Hawk. Contractor personnel, already highly trained, did a further 5-week course. The RAF did not. The Hawk T.1 Support Policy Statement said there was no requirement for pre-employment training for groundcrew before they worked at the Red Arrows, despite their procedures and practices being non-standard.

There inevitably came a time when no-one was properly trained. To help compensate, on-the-job training was carried out by Senior NCOs. But many had no qualifications as instructors, nor was there an approved syllabus. Consequently, as the Inquiry reported:

> '*A number of the six Red Arrows armourers had no recent experience of ejection seat equipped aircraft*'.

One had no seat experience in the six years since his basic training. That wasn't his fault, but neither was it Martin-Baker's. And no-one mentioned that, from the 1960s, the Fleet Air Arm and Martin-Baker combined to convert a Pusser's bus into a mobile classroom, touring air stations to train armourers on ejection seats. The bus included a ramp upon which a demonstration seat could be fired. The demise of such a valuable asset would have been a severe blow.

At the Inquest in 2014, one armourer was asked by Hugh Davies QC, for Lincolnshire Police, if he was aware that over-tightening the nut would prevent eventual release. He was not. But this admission was used to imply Martin-Baker had not provided the information, when in fact it was MoD who had not trained him. Exacerbating the deceit, Bernard Thorogood QC, for the Health and Safety Executive, told the court that the nut should be torqued down to 1.5 threads protruding; omitting that no trained fitter would contemplate such stupidity. This was no error. It was a lie.

These shortfalls were recorded as the top Risk to Life in the Engineering Risk Register. They had been highlighted during an Internal Quality Audit in January 2011 - Red Arrows' groundcrew were being *'left out on a limb'*. But what was omitted was even worse. These warnings had gone to the Chief of the Air Staff himself, along with a Business Case seeking £2.5M to correct immediate problems in the engineering management area of the Red Arrows. The author was told not to *'air Red Arrows' dirty laundry'*.

Again, this indicates proper application of risk management procedures at the working level, offering a clue where to begin... The same place Directors of Flight Safety aimed their own dire warnings throughout the so-called *'heyday of RAF Logistics Command'* and *'golden period'* of airworthiness.

Context

A brief discussion is necessary about the procurement strategy of the nut, bolt and Drogue Shackle Assembly; something I touched on before. It is MoD policy to appoint a single Design Authority responsible for the safety of a whole assembly, even if they do not supply each part. This is particularly important in a safety critical application. However, since the *savings at the expense of safety* policy this has been seen as an unnecessary expense, and is often only implemented if a knowledgeable project manager slips it into a contract.

Here, Martin-Baker did not supply the nut, having no control over what MoD bought. To compensate for such loss of control, someone must be appointed Co-ordinating Design Authority. Or, MoD must gamble and accept responsibility for a risk it has created. It is unclear how much MoD saved by adopting this strategy, but sometimes it is worth paying a few extra pennies for peace of mind.

*

A practical organisational problem Martin-Baker faced was, ironically, related to their long-term success. The sheer number of seat variants that remained in service, but which were considered obsolescent, required a huge, complex support infrastructure. Likewise MoD, who largely ignored legacy equipment, thinking it would take care of itself. It does not. It requires the retention of a trained and dedicated workforce - who know they are tainted by their experience, impeding advancement. It is telling that, of the two, only Martin-Baker could provide records from 1990. That alone should have been sufficient to dismiss the case.

Prior to 1993, MoD had many engineers who understood all the above, implicitly. After that date (when the RAF's policy to run down airworthiness management finally succeeded), MoD's corporate knowledge diminished rapidly. The Service Inquiry was thus hopelessly constrained before it even started.

*

Finally, it is worth spelling out MoD's formal position on information dissemination.

On 20 March 2018, Corporal Jonathan Bayliss, a Red Arrows maintainer, was killed at RAF Valley when flying as a passenger in Hawk XX204. A causal factor was that the pilot was unaware of new information regarding minimum altitude before commencing the fatal manoeuvre. This information was contained in a training manual that he was not required to sign for; although a higher level directive that he *did* sign for said he had to read it.

The Service Inquiry, and the subsequent Inquest in November 2021, both accepted that, notwithstanding the directive to read the training manual, he could not be expected to be aware of its contents because he did not have to sign to say he had. This was stated under oath by two RAF officers, and the Service Inquiry report signed-off by Director General Defence Safety Authority.

Now, if you will, apply this criteria to Sean's case. It is MoD's position that it was unnecessary to show the warning to those who had to be aware of it, because they did not have to sign to say they'd read it. That is, by MoD's rules there was no case to answer.

Have a little break now and think about that.

13: Airworthiness Review Teams, the Gas Shackle, and the 2002 QinetiQ report

Organisational failures

Airworthiness must be managed by 'Suitably Qualified and Experienced Persons'. Today, the military practice of equating Qualified and Experienced, with rank, prevails. There is no appetite to revert to the system whereby, regardless of grade or rank, integrity and competence are paramount, and staff assessed for these attributes. This would have eliminated one of the main root causes.

From 1990, repeated restructuring within MoD directly compromised airworthiness. Specialist staff recommended, for example, Training Needs Analyses be updated to reflect lower levels of expertise; in turn requiring Air Publications to be updated. But this, too, was shorn of funding. By 1992 there was little work being done, resulting in the many warnings from the Director of Flight Safety I have noted. Even if remaining staff had the expertise, there was no funding as the RAF supply organisation had robbed it to compensate for waste. As a consequence, Martin-Baker may also have been forced to dispense with their corporate knowledge, leaving them vulnerable to false allegations.

*

Once, in 1991, a Sea Harrier Board of Inquiry President, an RN Captain, called me to ask why the aircraft's microwave guidance system wasn't fitted, meaning the pilot couldn't get home safely after a bird strike damaged other sensors. My reply was - the RAF was in the process of scrapping this RN equipment, to avoid repair costs.

The Navy shuddered. I convened a meeting, where I demonstrated how it took five minutes to fit a Reynolds connector costing £5 to the Travelling Wave Tubes. I asked RAF suppliers why they had ordered millions of pounds worth of equipment to be scrapped for the sake of such simple repairs. (Verifying the repair cost a further £250. A new tube £15k. The Transponders and Interrogators it was fitted to, over £100k each. A Sea Harrier, at the time, excluding avionics, £16.2M. The pilot...). Answer - We can do what we want, the Navy has no say in it.

Flag Officer Naval Air Command staff were stunned, but now knew why only a third of their Sea Harrier fleet had a full navigation fit. To the RAF, the act of scrapping perfectly good RN kit removed the 'problem' of lack of funding to pay for its repair. That's one way of managing a budget. Another is to stop wasting money and make the aircraft safe. I concede many disagree. Thankfully, the pilot survived.

Airworthiness Review Teams

MoD routinely dissembles about these Reviews, carried out by the RAF Inspectorate of Flight Safety throughout the 1990s. (Causing much of the Nimrod Review to make little sense, as it assumed the Nimrod Airworthiness Review of 1998 was unique). For example, during the Mull of Kintyre Review (2010/11) the August 1992 Chinook Airworthiness Review Team report (CHART) was claimed to be only 52 pages long. However, the long-retired author came forward with information it ran to nearly 400 pages. Tellingly, the 'missing' pages contained perhaps the most embarrassing evidence imaginable. UK technical publications were so grossly out-of-date, RAF Chinooks were being maintained using Argentinian publications captured in 1982.[81]

Even after the complete report was revealed, MoD maintained it did not mention Chinook Mk2 (the variant which crashed on the Mull of Kintyre). A former RAF Chief Engineer (1991-96) repeated this to the media, as did Secretary of State Dr Liam Fox to the House.[82] In fact, it mentioned the Mk2 and its programme no less than 284 times. Significantly, it noted substantial cuts had been made to the airworthiness budget in the early 1990s. MoD has consistently denied having any record of this.[83]

Perhaps most pertinent of all, given the allegation against Martin-Baker, in 1995 the Director of Flight Safety warned the RAF Chief Engineer and Assistant Chief of the Air Staff that *'document integrity'* was routinely compromised at Air Stations and on deployments, because publications had been amended incorrectly.[84] He described the provision of information to operational units as *'inadequate'* and he was unable to *'establish the procedure by which these documents were kept amended'*.

This repeated warnings from his predecessor's Chinook report. And later, with increasing exasperation, his successor tried again in the 1998

---

81 Chinook Airworthiness Review Team report, Annex C, paragraph 4, August 1992.
82 Claim made by Air Chief Marshal Sir Michael Alcock (RAF Chief Engineer 1991-96).
83 Letter D/Min(AF)/AI MC06559/2006, 17 May 2007, from Adam Ingram MP, Minister for the Armed Forces to Steve Webb MP.
84 Tornado Airworthiness Review Team report, paragraph 127, 29 November 1995.

Nimrod report. Why were these senior officers (Air Commodores) ignored? The reason might be that their reports criticised Air Member Supply and Organisation, who had initiated the *savings at the expense of safety* policy in 1987. Also, the RAF Chief Engineer, who was required to take corrective action on airworthiness matters. In 1994 these posts were amalgamated, cancelling any possibility of criticism. The reports were not circulated or acted upon.

There is no excuse for not uncovering this during investigations. MoD's rules make this easy - the mandate that Technical Agencies are named in contracts, not referred to by post title. Their decisions form a significant part of the airworthiness audit trail, which MoD is required to retain for at least 15 years after the date of the last file entry.

The Gas Shackle

Figure 11 - Mk10A ejection seat with Gas Shackle. *(Public Domain)*

Martin-Baker introduced the Gas System Drogue Bridle release mechanism in 1984, as part of their process of continual product improvement. The primary reason for the modification was to eliminate a momentary geometric 'hang' during the disengagement process, in the process changing the Top Block. The design also served to eliminate the risk of shackle pinching by removing the Scissor Shackle. Consequently, Martin-Baker fitted it to all new seats, and since then no seat has been designed with a scissor. It recommended to customers that existing seats be retro-fitted.

Even when it belatedly adopted the design in Tornado, MoD did not fit it to Hawk - but it can be seen the status quo could not prevail, because a known risk (failure of parachute to deploy due to shackles not disengaging) was no longer ALARP. It can also be seen that this modification would have avoided the Top Block cracking that led to the illegal Routine Technical Instruction.

Martin-Baker were later accused by the Health and Safety Executive of not designing out the risk; one of many outright lies it perpetrated.

*

How does the modification proposal by Martin-Baker in 1984 link to the correct people in MoD knowing of the risk over-tightening? The overarching process for control and approval of modifications, and technical information flow, is (again) Post Design Services (see Chapter 5), with oversight provided by HQ Modification Committees.

In the first instance, Martin-Baker advised MoD that the Gas Shackle was now their Design Authority (and by definition the ALARP) Build Standard. We know this, because MoD rejected it. As part of the Investment Appraisal, the proposal would be cross-checked against, for example, the Risk Register. This would confirm a critical failure mode and Risk to Life could be eliminated, requiring Safety Arguments to be updated.

Hence, and even if we set aside all preceding evidence, MoD's aircraft and ejection seat offices (and British Aerospace) knew in 1984 of the risk of shackles failing to disengage, and why, at least six years before Martin-Baker's alleged failure to provide a warning.

QinetiQ report (2002)

QinetiQ is a defence company formed when the MoD's Defence Evaluation and Research Agency (DERA) was split in 2001.

Having determined that Tornado seats now had the Gas Shackle, the background was sought; particularly a report mentioned in a 2012 accident report. After many denials, MoD finally admitted on 2 January

2018 that it held QinetiQ report 'Mk10A Ejection Seat Modifications (02097 & 02198) for Tornado GR4/4A and F3 Aircraft - Phase 2', dated December 2002. Efforts to uncover more stalled when, on 8 January 2018, retired RAF engineering officer James Jones was declared *'vexatious'*, MoD refusing to provide further information. [85] (For 'vexatious', read *getting too close to the truth*). The penny had dropped - information helpful to Martin-Baker, but not disclosed to Inquiries or courts, had been inadvertently released. The shutters came down.

(The MoD correspondent, Mr Mark Bailey, was already acutely aware of MoD wrongdoing leading to a series of aircrew deaths, having been directly involved in advising Ministers to uphold the ruling that staff should be disciplined for refusing to make false record).

The 25,000 word report required careful analysis. In 1998 MoD had initiated a series of modifications, including the Gas Shackle previously rejected in 1984. It was necessary to establish how these changes to the Tornado Mk10A seat affected the Mk10B in Hawk. James and I hadn't quite finished at trial start date, when I was still expecting to be called at what was scheduled to be a six-week trial. But on the morning of 22 January 2018, Martin-Baker unexpectedly pleaded guilty to a single charge.

The report was a catch-up exercise, as MoD required independent safety assurance for the modification package. Trials had finished in 1999, but QinetiQ had not attended because they were not contracted to. Belatedly, this report completed the audit trail. The package was approved for Tornado in August 2007.[86] Retro-fitting was completed in 2010. The delays were not explained.

*

MoD had requested that seat operating limits be expanded to cater for heavier and lighter (female) aircrew. Withholding this detail meant most had inferred the *'shortcomings'* referred to were quality of design related, and a Martin-Baker liability. Had the XX177 Service Inquiry seen the report it would have had to say: *While the 1990-92 warning cannot be found, there is direct evidence of MoD's prior and subsequent knowledge of the risk, and its elimination by Martin-Baker.*

As the XX177 Inquiry overlapped that of Tornados ZD743 and ZD812

---

[85] Letter FOI2017/12801, 8 January 2018, from Mark Bailey, Defence Equipment & Support Secretariat.
[86] Tornado ZA554 Board of Inquiry report, paragraph 45.

(Moray Firth, three killed), and both considered Mk10 ejection seats, it is not unreasonable to expect the common issues to have been identified. Once again, the deception of MAA's supposed independence is exposed.

The QinetiQ report was copied to the ejection seat project team, who oversaw the submission of the modification proposals. Each project office (seat, Tornado, Hawk, etc.) now had to positively state their intent. Lacking this statement (in 'Cost and Brief Sheets'), the chair of the relevant HQ Modifications Committee is required to defer his decision. He cannot infer, not least because the validity of the Safety Case is at stake. He must be given a written assurance it is being updated. Certainly, he would query Tornado proceeding but Hawk not - because the decision to retain both designs would place an increased demand on the support infrastructure, which he is required to confirm is adequate.[87] Thus, a conscious decision was made not to modify Mk10B seats. It was not an error of omission.

MoD will not say why it did not modify Hawk, although Martin-Baker privately believe it funding related. (Noting that affordability is not an acceptable reason for not mitigating a risk). In reply to a Parliamentary Question, it merely briefed Ministers that it remained satisfied with the old design. [88] The Health and Safety Executive seemed not to understand that, to offer this vote of confidence, by definition MoD had to first understand the design and how to maintain it - destroying its case against Martin-Baker. But worse, both it and MoD avoided the fact that any Safety Case was no longer valid.

The effect on legal proceedings

The legal test for culpability requires harm to be 'reasonably foreseeable'. Martin-Baker *did* foresee the potential for harm and took all reasonable precautions. Initially by ensuring MoD received adequate information and training, ultimately by designing out the risk. The real question is - *Why did MoD ignore Martin-Baker?* Asked and answered.

*

The events I have outlined narrow down when and why the failure to reclassify the risk of shackle pinching occurred, and where one must

---

87 AUS(FS) 474, 1 January 1998. (Requirement Scrutiny Instructions).
88 Philip Dunne MP, Hansard, 10 and 25 February 2014.

look for the records of decisions. The suppression of these reports, and that it was members of the public who uncovered them, justifies the call for Service Inquiry reports to be made public *before* the Inquest.

It would be unfair to expect the police to identify such matters. But, once advised, they were obliged to review the evidence. They refused. After sentencing, on 26 February 2018 the Health and Safety Executive confirmed *it* was aware of the process I describe, but deemed it irrelevant. [89] Inconvenient to its case perhaps, but mandated airworthiness regulations are never irrelevant. The fragmented nature of its investigation, its malintent and incompetence, is laid bare.

*

On 24 January 2018 I sent the QinetiQ report and my comments to the Judge. She was now aware that further evidence existed, in MoD's hand, that it knew of but had decided not to eliminate the risk of over-tightening. Moreover, it had caused a Minister to mislead Parliament. This evidence confirmed:

- Senior MoD staff knew of the systemic failures but did not act.
- MoD failed in its duty to maintain an airworthiness audit trail.
- Martin-Baker communicated properly with MoD.
- There was no substance to the charge against Martin-Baker.
- The Health and Safety Executive knew all this.

After the sentencing hearing in February 2018, journalist Sean Maffett asked Martin-Baker's QC Richard Matthews if he had known about the QinetiQ report. He replied he had seen it, but didn't say when, or if he read it. I suspect he only saw it because the Judge sent it to the Health and Safety Executive on 24 January 2018, who in turn had to disclose it to Martin-Baker's solicitors. That is, *after* the guilty plea.

The Health and Safety Executive advised the Judge the report was irrelevant. (That word again). The truth was, initially (and to be kind), it did not understand it. Its thinking was one-dimensional. *It's a Tornado report, so not applicable to Hawk*. It continued in this simple-minded vein when claiming the parachute release mechanism in the Mk4 seat was different to that in the Mk10B, simply because they were different Marks. That's like saying the wheels on a Ford Cortina are round, so

---

[89] E-mail from the Health and Safety Executive principal investigator, 26 February 2018 11:43.

can't be round on a Ford Fiesta. At first, this may be ignorance. But when provided with proof, further denial was a lie. When repeated to the Judge, it formed the basis of her sentencing remarks, perverting the course of justice.

I simply note that an individual will commit an offence when he/she:

- Does an act or series of acts;
- Which has or have a tendency to pervert; and
- Which is or are intended to pervert;
- The course of public justice.[90]

The offence is aggravated if the offender involves others in the conduct (e.g. MoD), and evidence is concealed or destroyed (e.g. the exculpatory evidence).

---

90 R. v Vreones [1891] 1 Q.B. 360.

PART 4 - THE PURSUIT OF MARTIN-BAKER

'Fear the lords who are secret among us
Born of sloth and cowardice'
James Douglas Morrison

While individuals can be prosecuted under the Health and Safety at Work Act for failing to perform their prescribed duties, MoD itself cannot. At most it can receive a Crown Censure, by which the Health and Safety Executive records (in its opinion) there would be sufficient evidence to secure a Health and Safety conviction. Because this opinion is never tested in court, in practice it is an unfounded accusation. This encourages a high-handed manner and the ignoring of concepts such as evidence and truth.

A list of Crown Censures since 1999 can be found here:

http://www.hse.gov.uk/prosecutions/documents/crowncensures.htm

Those against MoD are worth a read. None relate to fatalities in aircraft accidents. On 26 February 2018 the Health and Safety Executive (HSE) was asked if it was considering a Crown Censure in this case, given MoD's admitted wrongdoing and the resulting death. It refused to reply, but when later pushed claimed it has no jurisdiction to investigate cases involving airworthiness failings. It stated that any complaint should be made directly to MoD, who would then judge its own case.[91]

This differs markedly to its decisions in respect of Nimrod XV230. Despite being formally asked to investigate MoD's failings on 15 November 2009, in October 2010 the HSE's Deputy Chief Executive, in a letter to Mr and Mrs Graham Knight who lost their son Ben in the accident, said that an investigation would be *'unlikely to obtain sufficient admissible evidence'*. This despite the Nimrod Review having accepted in full just such evidence.

He also stated that he would not investigate because there had been *'three previous investigations into the circumstances of the loss of XV230'*. On XX177 there had also been three, yet the HSE still conducted a fourth.

Finally, he claimed that bringing a prosecution could *'get in the way of*

---

[91] HSE letter 7 July 2020, from Jo Anderson, Engagement and Policy Division.

bringing in better practices'; completely missing the point that known senior staff had flatly refused to implement mandated regulations or meet their legal obligations.[92] That is, precisely the same violations were committed in each case.

Separately, that same year the HSE claimed passage of time (three years) as a reason not to investigate XV230. On XX177 five years had elapsed, but this time MoD was their collaborator.

To say these declarations cannot be reconciled is an understatement.

*

Having established that a known Risk to Life was not tolerable and ALARP, but declarations made that it was, let us look at the law:[93]

'Corporate Manslaughter is an offence created by Section 1 of the Corporate Manslaughter and Corporate Homicide Act 2007. It came into force on 6 April 2008. The offence was created to ensure that companies and other organisations can be held properly accountable for very serious failings resulting in death.

Corporate Manslaughter is wider in scope than the previous common law offence. There is a high threshold for liability, requiring proof of a gross breach of the relevant duty of care. However, it is no longer necessary to show that a person who was the "controlling mind" of the organisation was personally responsible for the offence.

Corporate Manslaughter relates to the way in which the relevant activity was managed or organised throughout the company or organisation. Wider considerations such as the overall management of health and safety, the selection and training of staff, the implementation of systems of working and the supervision of staff can be taken into account.

An organisation is not liable if the failings were exclusively at a junior level. The failings of senior management must have formed a substantial element in the breach. The prosecution must prove that the breach of duty was causative of death. The test is whether the breach made a more than minimal contribution to the death'.

I invite you to consider this against MoD's admitted failings. Especially given previous occurrences and the content of the Nimrod Review.

---

92 'The Crash of Nimrod XV230' (Trish Knight, 2012).
93 https://www.cps.gov.uk/legal-guidance/corporate-manslaughter

14: The Coroner's Inquest

A preliminary Inquest was held on 22 November 2011, and Pre-Inquest Hearings in 2013. The Inquest proper commenced on 9 January 2014, conducted by Stuart Fisher, Senior Coroner for Lincolnshire. He was required to determine cause of death, but his remit allowed him to explore events leading to the accident.

A feature of the proceedings was the robust defence of the seat design by Martin-Baker's Head of Engineering. When asked by the Coroner if he had read the Service Inquiry report, he replied that he had no need to. He repeated what the company had immediately announced - that their product was not to blame. This vigour upset some, particularly the Health and Safety Executive's QC. However, few if any understood his point. As MoD had a policy of servicing and modifying the seats itself, while ignoring Martin-Baker's instructions, then it forfeited the right to claim against or criticise the company. This principle is well understood in MoD, it being the main reason for Fault Investigation Requests being rejected by Technical Agencies.

As the Service Inquiry report had not been made public, only after media reports of proceedings could anyone offer comment. On 23 January 2014 I passed my initial assessment to the Cunningham family's QC, Tom Kark. It included a briefing on airworthiness failings I had recently been asked to submit to the House of Commons Defence Select Committee, by its chair James Arbuthnot MP. [94] Mr Kark replied *'evidence is finished'* and he could not raise the subject again. This emphasised the poor position his client was in due to MoD withholding the report. MoD's strategy of non-disclosure was working. Nevertheless, he was now aware the court and his client had been misled.

On 29 January 2014 Mr Fisher returned a narrative conclusion clearing Sean of blame, but was critical of Martin-Baker and MoD. On 26 February 2014, and as required by the Coroners and Justice Act 2009, he issued a report detailing his Matters of Concern. He listed only two, relating to the strapping-in process and Martin-Baker allegedly not providing information. Neither addressed the cause of death - a gross 'error' (more likely deliberate, to protect MoD) that the North Wales Coroner also made in 2021 in the case of Jon Bayliss. (A case in which at least 12 violations from the Cunningham case were repeated). He

---

94 Submission to House of Commons Defence Committee, 19 January 2014.

correctly mentioned the failure of other defences in depth, but did not make it clear he was criticising MoD - perhaps because he did not realise this. In August 2013 he had vowed to *'leave no stone unturned'*. But what if he is not shown the stones?

On 17 April 2014 Martin-Baker refuted these Matters of Concern, especially the characterisation of the ejection seat design as *'entirely useless'*. This prejudicial remark was widely reported, but Mr Fisher omitted that it was MoD who had rendered it unfit for purpose.[95] Due to the difficulty laypersons and legal authorities had understanding the technical and procedural complexities of the case, this constituted 'adverse publicity'. Martin-Baker were unfairly tried in the media, and the possibility of a future fair trial prejudiced. The point was later proved.

*

While making many recommendations, the Service Inquiry lacked context and few readers would understand what lay beneath. This meant, like the police and Crown Prosecution Service before him, Mr Fisher was prevented from reaching informed conclusions. The system is weighed heavily against bereaved families, and most of the evidence was a smokescreen thrown up by MoD and the Health and Safety Executive. Numerous improvements can be made; I have mentioned the main one - the Service Inquiry report should be independently analysed before the Inquest begins.

Lacking any examination of why the processes designed to produce a serviceable and airworthy aircraft were not implemented, Mr Fisher's conclusion was misguided. While the inadvertent ejection (cause of accident) and parachute failure (cause of death) were separate issues, they shared a common root cause - failure to apply MoD regulations, Martin-Baker's instructions, and training.

But in many respects Mr Fisher upheld the recent tradition of Coroners being robust with MoD. (One obvious exception being the 2003 Sea King ASaC Mk7 mid-air collision Inquest of January 2007, where the Coroner persistently prevented families from challenging untrue MoD statements).[96] Had the truth been told, he would have heard that major components of the Safety Management System were knowingly run down from the late 1980s. That, the failures were predictable, predicted,

---

95 http://www.bbc.co.uk/news/uk-england-lincolnshire-25943211
96 'Breaking the Military Covenant' (David Hill, 2018).

notified and ignored.

I am unsure if a Coroner should be expected to recognise this, and understand it wholly pollutes his Inquest. However, I have experience of other Inquests - for example, Nimrod XV230 and Hercules XV179 - where the Coroners' Officers and Investigators were pleased to take the same evidence from me and place it before the Oxford and Wiltshire Coroners, who acted upon it.

Lack of information from Martin-Baker was not a root cause or a factor of any kind, because MoD already it. Flat refusal to use it was. The Cunningham family left court under the misapprehension Martin-Baker had erred.

15: The Health and Safety Executive

'There are few things more dangerous than a mixture of power, arrogance and incompetence'.

Bob Herbert

The Health and Safety Executive (HSE) was given primacy after the Service Inquiry and 'investigations' by the civilian and military police were finished. (My only mention of the latter. They, too, are aware of the root causes, but refuse to act).[97] Those with a professional interest were in the dark between the accident (November 2011) and release of the Service Inquiry report (February 2014). Only the two Civil Aviation Authority directives, reflecting the Special Information Leaflets, offered clues.

Upon reading the Service Inquiry report I was immediately troubled. Root causes had not been addressed, and I sought to explain them in a short paper. I sent this to Martin-Baker on 11 March 2014, and updated it as more information emerged.

On 26 September 2016 the HSE announced its intention to prosecute Martin-Baker, despite MoD admitting it had *'habitually compromised'* safety critical activity.[98]

Two days later I sent my report to Helen Duggan, HSE media Communications Manager, asking her to pass it to investigators. She challenged my credentials, asking why I thought it should be entered as evidence.[99] I replied:

'The Hawk XX177 paper I sent you is one in a sequence of case studies. The failures are systemic, and what I describe on XX177 has been well-known since 1992, and fully accepted by Mr Haddon-Cave QC, Lord Philip, Coroners and RAF Directors of Flight Safety. Also, Government, when accepting the Nimrod and Chinook Reviews. One cannot look at these accidents in isolation, which I suspect is what is happening on XX177. Prior warnings of the underlying issues were given by those in senior airworthiness positions, and mitigation rejected

---

97 'The Crash of Nimrod XV230 - A Victim's Perspective' (Trish Knight, 2012).
98 Service Inquiry report, paragraph 1.4.5.20.
99 E-Mail from Health and Safety Executive Communications Manager Helen Duggan, 28 September 2016 16:23.

by senior MoD staff. There is an identifiable trend'.

No reply... Is it the role of a media office to vet potential expert witnesses?

I sent my paper to Lincolnshire Magistrates Court offices, for the (yet to be announced) presiding magistrate. Unconventional perhaps, but experience has shown me the effect of evidence being concealed. The legal system does not like having to regress. It would rather an injustice than admit to being misled. This has been demonstrated many times, most notoriously in the Chinook ZD576 case. One must find a way of getting the truth out there. And it is important I demonstrate here what the judiciary and HSE knew.

On 7 November 2016 I sent the HSE an update, emphasising Routine Technical Instruction/Hawk/059 was, according to MoD regulations, rogue. It could not fail to notice its allegations were fundamentally flawed, and I expected to be contacted. I heard nothing. On 16 January 2017 it confirmed it was prosecuting the company, but the details of the charge(s) were not revealed. I wrote again that day, asking if my evidence was thought, in some way, to be wrong.[100] No reply.

On 20 May 2017 I sought confirmation that my evidence had been seen by investigators. Again, nothing. I tried again on 26 October 2017. A reply on 30 October 2017 merely said: *'This e-mail is to notify you we have received your information'*.[101] Which does not mean it was read. Plainly, I was an inconvenient witness.

Evidence disclosure

Disclosure is the cornerstone of a fair trial. Material, either individually or collectively, must be disclosed if it *'might reasonably be considered capable of undermining the case for the prosecution or of assisting the case for the accused'.*

What may assist the accused must be objectively assessed from prosecution material, including what is said by the accused when questioned in interview by the HSE and/or police. By any reasonable interpretation my evidence met this test. The HSE was obliged to disclose it to Martin-Baker *'as soon as reasonably practicable after a not*

---

100 E-mail to Health and Safety Executive, 16 January 2017 17:36.
101 E-mail from Health and Safety Executive Communications Manager Natalie Dunn, 30 October 2017 14:40.

guilty plea in the magistrates' court'.[102] The plea was on 17 May 2017; the evidence was disclosed on 15 December 2017, only a few weeks before trial date.[103] The HSE had sat on it for 15 months, only reluctantly acknowledging receipt after many requests. By the time it was forced to disclose it, it had knowingly made false accusations against Martin-Baker, which were then repeated in court.

I was not the only one offering similar evidence, but the HSE didn't interview us. Yet it told the Judge and police it had conducted a full review. Both remain content at this lie. A layman might think the system that permits such disregard for the law defective. Notably, statute places a continuing duty of review on the prosecutor, even after sentencing. The HSE has not yet complied.

Tornado Airworthiness Review Team report (TART)

TART was issued in November 1995.[104] When asked for it in 2011, MoD denied having it, instead providing a list of its recommendations.[105] Yet recommendations are an integral part of all reports.

Recommendation #95 said:

> *'The Tornado Support Authority should ensure aircrew are kept abreast of the developments and initiatives in hand to overcome the shortcomings of the ejection seat'.*

Concerned that the recently completed Cunningham Inquest had not been told of this, on 10 February 2014 Angus Robertson MP asked:

> *'(1) What steps were taken in respect of recommendation 95 in the Tornado Airworthiness Review Team Report, and (2) What safety issues were identified in recommendation 95 regarding the Tornado's ejection seat?'*

Minister for Defence Procurement Philip Dunne MP replied on 25 February 2014:

> *'It is not possible today to re-identify the specific issue in the ejection seat that recommendation 95 was intended to address'.*

---

102 http://www.hse.gov.uk/enforce/enforcementguide/pretrial/after-disclosure.htm
103 E-mail from Health and Safety Executive, 7 February 2019 14:39. (Freedom of Information Request Reference 201902001).
104 Tornado Airworthiness Review Team report D/IFS(RAF)/125/74/1/1, 29 November 1995.
105 Letter 20111018-111845-007 from DE&S Policy Secretariat Air, 6 December 2011.

Encouraged (because you know you're on the right track when MoD lies to a Minister), re-assessment of the recently acquired Part 2 of the XX177 Service Inquiry, 'Schedule of matters not germane to the Inquiry', revealed:

'Item 94. British Aerospace Chief Engineer Review of Modification 02198 - Gas Shackle for Tornado'.

Why would the Service Inquiry dismiss such a safety modification as *'not germane'*, when it would have avoided the shackle pinching? Plainly, this was an attempt to conceal the fact that a modification was schemed for Tornado, but not Hawk.

Further investigations revealed that the aforementioned Service Inquiry into a Tornado mid-air collision on 3 July 2012 (ZD743 and ZD812, Moray Firth, Scotland, three killed), was not shown TART. With practiced guile, MoD said the information was *available* to the Tornado Panel. But it was not *made available*. Again, it took the public to identify the deceit.

It is important to understand who the recipients were - the RAF Chief Engineer and the Assistant Chief of the Air Staff. Both had key airworthiness roles. The former was responsible for correct implementation of regulations within his command - which by April 1991 included most of the airworthiness specialists I have mentioned. Instead, they were under orders to disobey regulations, but declare that they had been met. The latter signed and issued the Master Airworthiness Reference, declaring he was satisfied the Statement of Operating Intent and Usage, Build Standard and Safety Cases could be reconciled. Plainly, and as set out by the Service Inquiry, they could not. By 1995 this was a common failure across MoD, TART merely reiterating the situation.

By concealing TART, MoD misled everyone, by omission and commission. As ever there is a question of degree, but in 2012 (Service Inquiry) and 2014 (Inquest) MoD retained staff who were involved in the Mk10A seat modification programme (2007-2010). They were MoD's corporate knowledge. One was a Prosecution 'Subject Matter Expert' witness at the Inquest; and before that at an Inquest in October 2010. (See Chapter 24). The eventual release of TART meant it was now public knowledge the Coroner had been misled in 2014. MoD was required to advise him of this. It did not, and this (further) failure to disclose evidence was later prejudicial to Martin-Baker.

MoD finally yielded in May 2014, providing a heavily redacted version. It revealed the existence of more concealed evidence...

Ten key questions

When did the Health and Safety Executive know:

1. Instructions had been issued by the RAF contrary to those provided by Martin-Baker, and the ejection seat had not been certified serviceable.
2. RAF maintainers at other stations were trained differently.
3. MoD regulations required a new stiffnut and bolt each time.
4. Incorrect tools were used, despite the correct ones being supplied.
5. MoD had signed the report arising from customer queries.
6. A modification that eliminated shackle pinching was available in 1984, but rejected by MoD. And that Martin-Baker had fitted this to all Mk10 seats delivered to other customers since 1990.
7. That this modification was retro-fitted to Tornado in 2007, but not Hawk, meaning the risk of shackle pinching was not tolerable and ALARP
8. A modification, in the form of a shroud, was available in early-1980s to mitigate misrouting of seat strap. And, that on a number of occasions MoD rejected this in all aircraft except Harrier.
9. The escape system in Hawk was not tolerable and ALARP.
10. MoD had repeated the offences identified by the Nimrod Review.

The answer to each is - *Before the trial of Martin-Baker*. Its proven incompetence had now drifted into outright abuse of entrusted power, significantly subverting the legal system. Otherwise known as systemic corruption.

16: Lincolnshire Police & the Crown Prosecution Service

Lincolnshire Police

'Inquiry' and 'investigation' can be used in lieu of each other, although there are subtle differences. In this context an Inquiry is a legal process. I may not conduct an Inquiry, but I may investigate.

If the police decide to investigate potential criminal activity or respond to a complaint (neither of which they are obliged to do), they should gather evidence which either indicates or helps to explain contributory circumstances. An assessment is submitted to the Crown Prosecution Service for a decision on whether to prosecute. Fundamental to the judicial process, <u>all</u> evidence must be presented, to enable an informed decision. Evidence is defined as *anything material that tends to prove or disprove the case.*

As a person had died within its jurisdiction, Lincolnshire Police conducted an investigation - which they later distanced themselves from, explained in Chapter 23. Nevertheless, initially there was honest endeavour. They had to consider sabotage, and given the degree of over-tightening of the bolt this was their initial concern. A few days after the accident, the seat was taken to a classroom at Martin-Baker's factory, and the room declared a crime scene while the investigators and MoD personnel were taken through the design and operation of the seat by a former RAF armourer. (Who ably proved his training was up to the task, which should have made the police and the Health and Safety Executive ask who taught him).

The investigation lasted around nine months (deduced from the Service Inquiry being convened on 12 September 2012), but it remains unclear what evidence was taken. As the police now deny they had primacy, and will not even acknowledge if a report was written, assessment is impossible.

Lacking the expertise to assess the procedural and technical matters raised in a typical MoD accident report, far less work out what has been omitted and concealed, the police must be regarded as incapable. That is not a criticism, just fact. They need independent assistance, but that is nigh on impossible if MoD withholds its reports. So, expert analysis was only possible after the Inquest ended in February 2014. Nevertheless, they seriously erred by accepting the MoD report at face value, when common sense, and the blatant covering up of past fatal accidents, meant it should have been regarded as tainted evidence. More so, given the Convening/Issuing Authority was directly implicated, and the Nimrod Review was fresh in everyone's minds.

The police did not follow the evidence, allowing MoD to determine the scope and direction of the investigation. This is common (mal)practice, the Nimrod XV230 case being a prime example. (Whereby Thames Valley Police told bereaved families that they had decided not to investigate as it was unlikely witnesses would come forward. At the time, they were sitting on witness statements).

*

All police stations have 'Collators' (or used to). Officers brought him information, which he used to identify linkages. Today, this is computerised - the Police National Database - although clearly the person interrogating the database must know what to ask. Lincolnshire should have been able to type in 'military aircraft accident' and gain access to (e.g.) Thames Valley Police records on Nimrod XV230 and Strathclyde Police's on Chinook ZD576. The similarities would stand out like a sore thumb. Forty-three deaths, caused by the same failures.

Immediately, the accident should have been viewed as a recurrence, with MoD the prime suspect. But the investigation, like MoD's and later the Health and Safety Executive's, was too narrow and shallow, ignoring trends and detail. Nevertheless, unlawful activity was suspected (because MoD admitted it), and charges against Martin-Baker and (the wrong) MoD individuals considered. The inclusion of Martin-Baker was the give-away, revealing poor understanding.

Later, the police were given direct evidence that they had been misled by omission and commission, but did not pursue it. Therefore, no Inquiry or investigation has been completed, because the truth has been concealed, facts ignored, and evidence not reviewed.

Crown Prosecution Service (CPS)

In March 2013, two months before the Service Inquiry was signed off, Lincolnshire Police passed their findings to the CPS. On 13 April 2013 Alison Storey, a prosecutor for its Special Crime and Counter Terrorism Division, issued this statement:

'I have concluded there is insufficient evidence to establish any individual or the MoD breached their duty of care to Flight Lieutenant Cunningham in

relation to the servicing of the relevant equipment. I also concluded there is insufficient evidence to provide a realistic prospect of conviction of Martin-Baker in relation to the manufacture of the parachute. I have therefore concluded, in accordance with Code for Crown Prosecutors, no charges should be brought in connection with the death of Flight Lieutenant Cunningham'.

Like the police, the evidence she assessed did not include independent expert analysis of MoD's report, which was only partially released to the public 11 months later. This report was, in effect, a confession. Ms Storey must have been reading something else to claim there was insufficient evidence. Her narrow focus on the servicing of the seat and manufacture of the parachute was inept - particularly mention of the latter, as Martin-Baker do not make it.

Perhaps it would be wise to have someone who has read the evidence make these decisions? Like the police, the CPS was aware of the systemic failings I describe. No-one could fail to notice Mr Haddon-Cave had reiterated the same failings in 2009.[106]

I conclude that the CPS (and later the Health and Safety Executive) wilfully ignored notifications of MoD's wrongdoing, and the common factors set out in the Nimrod Review. This subversion placed current and future aircrew at greater risk.

Failure to investigate and concealment of evidence

To determine whether material or enquiries are relevant to the investigation, investigators need to ask: *Does this have the capacity to impact on the case?* What advice did they seek to make the determination of relevance? The police investigation is a closed book, but the Health and Safety Executive's *'star witness'* (its words, to the BBC on 30 January 2018) was a Military Air Accident Investigation Branch officer involved in the investigation. I am unconvinced as to the probity of this, given the Military Aviation Authority's direct involvement. It smacks of collusion, and treating MoD as a co-prosecutor.

Misled, the police did not understand the contributory factors and events. When given evidence on root causes they ignored it, failing to interview witnesses. The CPS probably spotted a huge minefield, involving misconduct by persons in public office. It decided not to proceed, but its reasoning was vacuous rubbish. Throughout, the Health

---

106 Letters to Crown Prosecution Service, 29 September 2012 and 4 December 2014.

and Safety Executive was working alongside the MoD, police and CPS. It developed misconceptions which it later relied on in court. And persisted when shown evidence they were flawed.

After being cleared, MoD admitted liability and offered a paltry financial settlement to the Cunningham family, subject to them signing a gagging order. Knowledgeable observers began to suspect another cover-up.

At this point, Martin-Baker were not accused of anything. Did the company think to offer the Cunningham family or their legal representatives a briefing? *Here's what we recommended MoD do, and here's what its regulations require it to do. And this is what it actually did...*

17: Court Appearances (2017)

Martin-Baker appeared in Lincoln Magistrate's Court on 16 January 2017, entering no plea. On 17 May 2017 they pleaded Not Guilty. Trial was set for 22 January 2018 at the Crown Court. It was to be heard by Justice Charles Haddon-Cave, author of the Nimrod Review, but instead he was allocated the Parsons Green bombing case. A pity. As soon as he read MoD's admission that it could not find a Safety Case Report, his experiences on Nimrod would come into play. Did he recuse himself? Having devoted his 2009 report to the lack of a valid Nimrod Safety Case, and demanding wholesale changes in MoD, here was evidence of even worse failures. No-one could wish for a better defence witness.

*

Martin-Baker were instructed not to discuss any aspect of the case, not even to potential defence witnesses.[107] It was only in October 2017 that their solicitors, Clyde & Co., asked to interview me regarding the evidence I had submitted in 2014. It lasted five hours. I was accompanied by retired navigator Squadron Leader Sean Maffett, now a respected aviation journalist, who confirmed his own training in 1958 included seat operation.[108]

I offered my thoughts on Justice Haddon-Cave's involvement. Clyde professed ignorance of the Nimrod Review - distinctly odd for a specialist Health and Safety firm. They dismissed MoD's inability to find information as a misunderstanding; more concerned with defending the Seat Pan Firing Handle (quality of product) accusation, which at the time was still on the table.

I disagreed. The quality accusation was easily refuted (and so it proved). Less so the information one, as it required detailed knowledge of MoD's ever-changing structure and procedures, over several decades. Also, someone to explain, potentially to a non-technical jury, and certainly a Judge, the engineering principles and MoD procedures I have set out. That would inevitably require examination of MoD maintenance policy decisions, knowing where to look for historical records, and who to call as witnesses. But at a more basic level my reasoning was straight forward - MoD had declared itself content with the seat design, but lied over the

---

107 Telecon Hill/Andrew Martin, 21 March 2017.
108 Clyde & Co. Solicitors (Mr Roger Cartwright), 31 October 2017.

information. Always follow the lie.

I was shocked by the reply. The proposed strategy (yet to be put to Martin-Baker) was *'not to highlight MoD's failings'*. That is, not mention they were recurring and in recent years had killed scores of airmen and passengers. MoD's guilt was to be brushed under the carpet, at the expense of Martin-Baker's reputation. But I understood their concern. MoD is notorious for threatening contractors (and its own staff) if they question wrongdoing. As the public is prohibited from complaining about maladministration in the Civil Service, I am unsure what a Judge's reaction would be to using this in mitigation.[109]

Nevertheless, I was asked if I would track down corroborating witnesses. (Presumably in case Martin-Baker rejected the strategy). Clyde wanted someone senior from MoD's Aircraft Technical Publications department in 1990, and someone trained on ejection seats. On 9 November 2017 I advised them of two retired civil servants who would confirm my evidence on airworthiness and publication procedures. Similarly, on 17 November of a retired RN Air Engineering Officer who (in addition to Sean Maffett) would describe the training he received before 1990. So, all bases were covered - procedures, publications, training, aircrew and engineering - each by more than one witness, and each describing the period *before* the alleged offence.

Clyde acknowledged receipt, then clammed up. Both strange and rude given my assistance. After all, it was *their* job to find witnesses and mount a defence. I felt the lateness of evidence gathering indicated complacency, and a dangerous naïvety in assuming MoD would heed disclosure rules. Other warning signs were evident. That same month MoD had publicly rehearsed its position, with a former Hawk Type Airworthiness Authority giving lectures about the accident.[110] This free shot seemed particularly inappropriate given Martin-Baker were told not to speak to witnesses. Later, the Judge noted Martin-Baker had commented to the media, and were possibly in contempt. She did not mention MoD's abuse of its position.

*

---

109 Letter AP000060 from Civil Service Commission, 1 May 2012.
110 https://dev.aerosociety.com/events-calendar/xx177-hibernations-in-the-machine-introducing-the-pressure-release-model/ The Type Airworthiness Authority is responsible for the airworthiness of the aircraft fleet, whereas the Continuing Airworthiness Management Organisation is responsible for individual aircraft.

Shortly before trial Martin-Baker changed their plea, admitting a single charge of:

'Failing to ensure the safety of non-employees, contrary to Section 39(1) of the Health and Safety at Work etc. Act 1974, and Section 33(1)(a) of the said Act'.

In the intervening two months a corporate decision had been made to roll over. Just take the hit for MoD. The path was clear for it to continue offending. It rushed down it.

'A dreadful position'

Six years after the accident, and having presumably carried out their own investigation, Martin-Baker and Clyde (correctly, if politely) thought the information issue a *'misunderstanding'*. Yet, in the same breath Clyde advised their client to plead guilty, and Martin-Baker did so.

It is worth noting that the legal maxim of 'innocent until proven guilty' does not always apply here. In a prosecution under Health and Safety at Work Act, it is for the Defence to prove, on the balance of probabilities, that it was not reasonably practicable to have done more than was done to satisfy its duty.

In July 2017, HM Inspectorate of Constabulary and the Crown Prosecution Service Inspectorate had accused the police and Crown Prosecution Service of failing to disclose crucial information about cases, undermining the right to a fair trial. The Criminal Law Solicitors' Association said clients could be put in the *'dreadful position'* of being advised to plead guilty to receive maximum credit without all the evidence having been presented to the court.[111]

Precisely what happened in this case.

---

111 https://www.independent.co.uk/news/uk/home-news/Police-crown-prosecution-service-disclosure-lawyers-trial-a7846021.html

18: Trial Day, 22 January 2018

The trial opened with an announcement of the change of plea. To what, precisely, was still uncertain, as the charge was a bit of a catch-all. Even the Health and Safety Executive was unsure - or at least said it was, to journalists. The Judge, Mrs Justice Carr, was not amused, not least because this unexpected turn of events disrupted the work of the entire Midland Circuit. Scheduled for six weeks, proceedings were brief.

Many potential witnesses, who had been waiting for some months, realised they would not be called. Others were still awaiting initial contact. They were not informed by Clyde & Co. and had to rely on media reporting.

Danny Savage, the senior BBC correspondent covering the trial, e-mailed me that lunch-time:

'Very little was mentioned in court apart from documents which weren't read out! It's a repeat of what was said at the Inquest'.

In other words, the Prosecution didn't disclose the exculpatory evidence. But neither did the Defence.

Danny called the following day. Like most journalists he had accepted MoD and Health and Safety Executive claims as true, unaware MoD already knew the risk of over-tightening the nut had been eliminated in 1984. I again provided evidence they were false. But news stories come and go quickly, and he and his colleagues Louise Wheeler and Gemma Dawson were reassigned, promising to return at the sentencing hearing; which the Judge set to commence on 12 February 2018.

Why change plea?

Martin-Baker's annual report of 7 November 2017 stated:

'On the evidence presented by the Prosecution as at 11 October 2017, it is the legal opinion of both Clyde & Co. and Defence Counsel there is a realistic prospect of the Company successfully defending itself at trial and it is probable i.e. more likely than not, the Company will be successful at trial. A not guilty plea has been entered and the matter is being fully prepared for trial'.

This contradicted Clyde's statement of 27 October 2017 - the intention to advise Martin-Baker to plead guilty. Yet, it aligns with an e-mail they sent me on 17 November 2017:

'At this point, we are continuing to prepare our client's best case for the trial in January 2018'.

Our last correspondence before trial was on 20 November 2017, when (at their request) I provided further details of witnesses. I inferred Clyde had not yet put their proposal to their client, or it had been rejected. What happened between then and the trial date to cause company director James Martin to change plea? He didn't decide on a whim at breakfast on 22 January 2018. There would have been board meetings. Meetings with Clyde, Defence Counsel and insurers. It wasn't as if fresh evidence had emerged indicating guilt. Quite the contrary. The QinetiQ report of 2002 had been obtained under Freedom of Information in early January. The Health and Safety Executive, Clyde and the Judge had it. All knew, with absolute certainty, the charge was without merit.

Undoubtedly there were machinations and arguments, both within Martin-Baker and between them and the MoD/Prosecution. Andrew Martin, Vice President of Business Development and Marketing, had vowed to fight to the bitter end. His position prevailed for all but the last few days before trial. It must have been dispiriting listening to the false accusations and downright lies, which were blithely repeated, accepted by legal authorities, and reported by the media, knowing they were wrong.

Observers considered a number of factors. A compulsory purchase order was being mooted on the company's Chalgrove site in Oxfordshire (formerly RAF Chalgrove), where its extensive airborne testing facilities are. MoD had transferred ownership of the site to the Homes and Communities Agency in 2016 (now Homes England), who promptly announced plans to build on it. Not long after sentencing, the media reported the threat was being withdrawn. Also, large orders were imminent from Israel, Japan and South Korea. The company's Australian arm was going through the approved maintenance organisation process, granted on 23 April 2018. These commercial considerations, involving potential income in the hundreds of millions, cannot be ignored when the maximum fine was 'only' £1.6M.

Ultimately, and despite what Clyde & Co. said in October 2017, the decision was made in order to clear any perceived obstacle to a forthcoming tender for a US contract, who are exploring next-generation seats. (To Martin-Baker the MoD is a minor customer, although in reputational terms it remains beneficial for them to be able to say the RAF use their seats). At a purely business level this worked. In the US, there would be vague awareness that a finger was being pointed. Getting it out of the way, and avoiding a lengthy trial and more publicity, especially when the media and legal authorities were not interested in the truth, was expedient.

But the decision had unintended, although easily foreseen, consequences. It hid MoD's serial wrongdoing, leading to further deaths. Martin-Baker must share culpability for these; but of course not to the same degree as MoD, the Health and Safety Executive, police and the judiciary.

19: Between trial & sentencing

On 24 January 2018, two days after the trial opened and closed, I submitted my evidence to Rebecca Collins, clerk to Mrs Justice Carr. Primarily, this was the report I had sent to the Coroner, police, Clyde & Co., and the Health and Safety Executive, now updated with my initial assessment of the QinetiQ report. To recap, this confirmed that in 1984 MoD knew of the risk of shackle pinching, and the risk eliminated by introducing the Gas Shackle. I thought this one of several silver bullets.

Ms Collins confirmed receipt, saying my papers had been passed to the Judge who had instructed her to:

'Forward (my) e-mail and attachments to the Prosecution so that the Prosecution can consider any questions of disclosure and, if appropriate, provide the Judge with copies of any material also provided to the Defence'.[112]

At the same time, my friend James Jones advised the Health and Safety Executive of the 2002 QinetiQ report.[113] Principal investigator David Butter asked for a copy. We inferred MoD had not disclosed it. It was sent on 26 January 2018.

*

On 27 January 2018 I wrote again to Rory Stewart MP, formerly chair of the Defence Select Committee (and aware of all the issues I have raised here), who had recently been appointed to the Ministry of Justice. To avoid repetition, I précised by emphasising the same failings had been admitted by MoD in other fatal accidents. His Customer Investigations Team replied on 5 March, saying the Ministry could not comment on or intervene in judicial matters - the Judge's decision was final. In reply, I pointed out Mrs Justice Carr had been systematically misled and an offence had been committed by the failure to disclose evidence. I asked: *'Whose role is it to oversee judicial fairness?'*. No reply. I'm sure the Ministry can justify its words in a strictly legal sense, but to a layman this was the Establishment closing ranks.

On 6 February 2018 I forwarded my expanded evidence to the Coroner, Stuart Fisher, copied to the Judge's clerk, Martin-Baker and Clyde & Co. Mr Fisher now knew his court had been misled in 2014. He replied the same day, confirming he had immediately forwarded it to Mrs Justice

---
112 E-mail from Judge's clerk 26 January 2018, 12:59.
113 E-mail to Health and Safety Executive Media Inquiries, 23 January 2018 15:10.

Carr.[114]

Importantly (I thought), evidence had now been sent to her from a judicial source. The guilty plea meant she was having to rely on what was said in the Coroner's Court in 2014. It might be thought she would be keen to understand these new papers proving that evidence false. Even a brief glance at the covering letter should have alarmed her. Here was a member of the public providing evidence that MoD denied the existence of, which wholly refuted the Prosecution's allegations.

I could only wait, but naturally hoped she would hand down a sentence reflecting the fact Martin-Baker had pleaded guilty to put the matter to rest, taking the hit for MoD. But the Health and Safety Executive advised her this evidence (proving their allegations baseless) was irrelevant. Legal advice to me is that a Judge is not permitted to bang heads together and say *You're wrong, I have evidence before me proving it is not only relevant, but exculpatory*. But one way or another justice demands it be said, because not doing so places aircrew, groundcrew, passengers and the public at greater risk.

The charge, and withdrawal of the design defect allegation

On 20 February 2018, the HSE issued this:

STATEMENT OF OFFENCE

FAILING TO ENSURE THE SAFETY OF NON-EMPLOYEES, contrary to Section 39(1) of the Health and Safety at Work etc. Act 1974, and Section 33(1)(a) of the said Act.

PARTICULARS OF OFFENCE

MARTIN BAKER AIRCRAFT COMPANY LIMITED, being an employer, on and before 8 November 2011, failed to conduct its undertaking, namely the design, manufacture, supply and support of the Mark 10B ejection seat system, in such a way as to ensure, so far as was reasonably practicable, that persons not in its employment who may have been affected thereby, including Flight Lieutenant Sean Cunningham, were not thereby exposed to risks to their health or safety, in particular by non-deployment of the main parachute attached to the Mark 10B ejection seat during low speed or zero-zero ejections.

This did not reveal much, except confirm the focus was on the pinched shackles. Hitherto, it was widely assumed the allegations related to the

---

114 E-mail from Lincolnshire Coroner Stuart Fisher, 6 February 2018 14:44.

Coroner's Matters of Concern, including strapping-in. The HSE later expanded:

> 'In November 2017 the HSE confirmed that the inadvertent ejection [cause of the accident] *was not caused by any fault attributable to the company*. Upon receiving clarification of the HSE's case, the company accepted a breach of Section 3(1) of the Health and Safety at Work Act, on the basis that it failed to provide a written warning to the RAF not to over-tighten the drogue shackle bolt'.[115]

It is important to understand when to use 'RAF' and when to use 'MoD'. The HSE (and others) consistently got it wrong. The information was provided, as required, to MoD and its predecessors. It was an internal MoD function to disseminate it to the RAF. I'm not being pedantic. If you ask the wrong person, don't expect an accurate answer. Few in the RAF are required to know what I've set out here. But all civilian engineers in MoD aircraft project offices are.

*

In a statement issued after sentencing the HSE said:

> 'Regardless of whether Sean needed to exit the aircraft or not, he still should have survived. The HSE didn't look at what initiated [the ejection] because had he needed to use it, the outcome would have been the same'.[116]

This was an outright lie. The HSE did look at what caused the ejection - and raised it again in court. In October 2016, its Initial Case Summary had alleged the Seat Pan Firing Handle safety pin design was defective. That, if it could be shown a modification to the handle could have prevented ejection, then Martin-Baker were guilty of negligence for not having taken such action. This overlooked that Martin-Baker had taken action, but in the 1980s MoD repeatedly rejected a modification to fit a shroud to the Handle, preventing strap misrouting. This was belatedly introduced after the accident, MoD implying *it* had taken the initiative.

It also ignored that there quickly comes a point when the contractor cannot just change the design used by MoD. It may develop and propose a modification, but cannot dictate uptake. A series of Design Reviews present the proposed design and, ultimately, the product, to MoD - who

---

115 https://www.blstraining.co.uk/martin-baker-fined-1-1m-red-arrows-pilots-fatal-ejection-seat-failure/
116 https://www.ioshmagazine.com/article/martin-baker-fined-ps11m-red-arrows-pilots-fatal-ejection-seat-failure

formally accept it. At every stage operators and maintainers are invited. Over 7,600 successful ejections speak to quality of design.

Minds locked shut

So, the final charge related only to the supply of information. I have set out the evidence proving MoD not only had the necessary information, and issued instructions not to use it, but also issued contrary instructions (e.g. to torque the nut, and omit disturbed systems testing).

More emerged in correspondence with Mr Butter. When I put the evidence to him again - witness testimony and MoD papers confirming the information had been regularly provided since the 1950s - on 26 February 2018 he replied saying it was *'not accepted'* that MoD knew of the risk of over-tightening the nut before 1990. That, it was never aware, until after the accident. He added:

> *'Both parties were in possession of the information you refer to, and neither considered it relevant'.*[117]

In fact, the Defence had not commented. Besides, the Prosecution does not speak for the Defence. The information he referred to included the *intended* evidence of witnesses who had presented themselves to Clyde & Co., but had not been interviewed. This implies Clyde advised the HSE of what they *thought* the witnesses would say if interviewed. Evidence that has not been taken or presented cannot be challenged, so the court was misled based on hearsay and false assumption. The HSE later admitted it did not review my evidence, and had none supporting its own proposition.

More fundamentally, Mr Butter ignored that Martin-Baker's alleged administrative oversight had no bearing whatsoever on Sean's death. There was no causal link (a relationship between alleged conduct and outcome), primarily because the RAF had chosen to disregard Martin-Baker's instructions, and continued to do so even after the trial. On the other hand, there was a causal link to MoD's admitted misconduct.

Thus, the entire basis of the charge was wrong.

*

The HSE dwelt at length on the fact five users - India, Pakistan, Egypt,

---

[117] E-mail from Health and Safety Executive principal investigator, 26 February 2018 11:43.

Italy and Finland - had received notifications between October 1990 and July 1992 of the risk of shackle pinching. In a magazine interview after sentencing it said:

> 'In light of...the overseas manuals, it was our view Martin-Baker had failed to adequately control the risk of an <u>interference fit</u>'.[118]

(An interference fit is normally a deliberate design feature, typical examples being wheel bearings or when a joint must be waterproof. To remove all ambiguity, here the shackles were forcibly jammed together due to inappropriate and careless over-tightening).

Hence, Martin-Baker manuals are central to its case. They warned:

> 'Observe that the mechanism has operated and has removed restraint from the Scissor, thereby freeing the Drogue Shackle'.[119]

In other words, if the Drogue Shackle does not disengage, one has to ensure it does. If the correct parts are used, and are serviceable, then the only reason can be an over-tightened nut. Immediately, the focus of any investigation should have been on why the <u>RAF</u> ignored this check.

The truth is that Martin-Baker *had* controlled the risk, ultimately by designing a new Gas Shackle system in 1984. MoD failed to control it by not ensuring serviceability, and declining to embody the Gas Shackle in Hawk seats - but falsely stated it *had* been controlled. The HSE knew this before the article was published, confirming receipt of the evidence on 29 January 2018.[120] As stated, I had sent this to the Judge, who also passed it to the HSE. The HSE was required to notify the Judge of the truth. Instead, she was informed this evidence was irrelevant; later relying on this disinformation when determining punishment.

Moreover, regarding the mechanical shackle assembly, the HSE suggested that issuing maintenance guidance alone was insufficient. That, a different design was necessary. The Judge disagreed:

> 'As for the original design of the shackle assembly, it was designed in 1949. The prosecution experts have not addressed its adequacy explicitly by reference to standards in those days'.[121]

The important part is that the release mechanism design dates to 1949.

---

118 https://www.ioshmagazine.com/article/martin-baker-fined-ps11m-red-arrows-pilots-fatal-ejection-seat-failure
119 Martin-Baker Instruction and Servicing Manual M-B/17, paragraph 107b.
120 E-mail from Health and Safety Executive, 29 January 2018 12:01.
121 Sentencing remarks of Mrs Justice Carr, 23 February 2018.

The HSE later disagreed with the Judge, denying any relationship between this and the Hawk Mk10B seat.[122] One can only speculate as to why it did not raise this objection in court.

*

Having refused to interview witnesses who came forward with direct evidence, who did the HSE speak to? It was suggested to me that anyone wishing to know about Martin-Baker seats, but without speaking to the company, would engage Survival Equipment Services Limited (SES) of Tetbury. This proved accurate, and I asked if they would discuss the matter. They politely declined, citing a confidentiality agreement with the HSE. This relationship has never been made public, so what the HSE asked is not known. But the managing director did say:

> 'Whilst we appreciate your concerns, and we at SES have ours, they are not for public discussion'.[123]

Perhaps not by SES, but they certainly need to be made public, because the safety of the public whom military aircraft overfly forms part of the definition of airworthiness, so is a matter of public interest.

*

On 26 September 2016 the HSE claimed in a press release: *'We have conducted a thorough investigation'*.[124] Really? It had a duty to seek, obtain and understand evidence. It did not, failing to uncover or later heed expert testimony. Most obviously, any competent investigator would surely twitch at MoD's claim not to understand the most basic feature of a design that has been in service for 60 years. (A claim so ludicrous it is beyond parody). And, plainly, it did not grasp that one cannot separate the acts of fitting the nut and bolt, and checking shackle disengagement. One follows the other, as night follows day.

In reality, the HSE investigation was cursory, incompetent and venal, carried out with a closed mind.

---

122 E-mail from Health and Safety Executive principal investigator, 17 May 2018 12:36.
123 E-mail to author, 5 March 2018 13:47, from SES Ltd, Tetbury.
124 https://www.bbc.co.uk/news/uk-england-lincolnshire-37475298

20: Sentencing hearings, 12/13/23 February 2018

Day 1

Mrs Justice Carr had received a letter that morning. She passed it to the Prosecution, who passed it to the Defence, who asked for time to digest it. The contents were not revealed, except it contained *'significant'* information of events from many years ago - indicating she had read it. Yet, she claimed not to have read third-party correspondence, implying who wrote it was her main consideration. That can only have been MoD. After a 25-minute recess, she stated the hearing would finish the following day, and would reserve her judgment until 23 February.

*

Rex Tedd QC, prosecuting, attacked Martin-Baker on points MoD knew to be misleading and untrue. He alleged Martin-Baker had exposed users to risk by not ensuring MoD knew how to service the seat, breaching its duty of care. He did not refer to the evidence, in MoD's hand, or to the witnesses who had come forward, proving otherwise.

He made much of the 1971 contract to develop variants of the Mk10 seat for MoD; that it *'shall be capable of safe escape at all heights and speeds'*. But he omitted that such a design aim cannot be an absolute requirement. It carries conditions, such as *subject to the seat being correctly maintained by trained staff*. He omitted that the Mk10 seat was capable, MoD had accepted this, and had been proven in many ejections.

Former military accident investigator Lieutenant Commander Graham Hamilton RN, acting for the Prosecution, demonstrated the Shackle Assembly to the court. The Judge tried it out, but no mention was made that the previous certification of serviceability was void. Separately, outside the court, Sean Maffett asked Hamilton if he was aware of the Gas Shackle, and that it would have prevented the pinching. He would not comment, saying such matters were above his pay grade. (So why was he a witness, if he could not comment on something so basic?). He also characterised as *'rubbish'* the suggestion that the Release to Service (the Master Airworthiness Reference) must be underpinned by a valid Safety Case.[125] It was clear from his words that he had not actually checked if a Safety Case existed; entirely negating his investigation.

Moving on to the Drogue Shackle nut and bolt, Mr Tedd persisted with

---

125 E-mail Maffett/Hill, 13 February 2018 17:56.

the *1.5 threads* line, despite his client knowing this was contradicted by MoD publications, trade training, and good engineering practice. He also overlooked that the matter had been laid to rest at the 2014 Inquest (see Chapter 9). He implied *1.5 threads* was the final test of safe functionality, taking up much time and obscuring the certification issue.

It was alleged no drawings existed of the Scissor and Drogue Shackles when linked. This was misleading, as technical publications show the shackles linked and describe their operation, emphasising *'ensure the system functions correctly'*.[126] Once again the Health and Safety Executive sought to divorce the act of disassembling and reassembling the Drogue Shackle from the need to check serviceability. This is akin to a garage not checking wheel nuts after changing a wheel.

Martin-Baker were accused of having *'little appreciation of the effect of tightening down on the nut'*. As the words were spoken in the 'ordinary course of legal proceedings', strictly speaking they were not defamatory - but they were patently untrue, witness the servicing instructions and the Gas Shackle eliminating the risk of over-tightening. And, that it was Martin-Baker staff who immediately identified the circular rub mark on the Scissor Shackle; which would have confirmed to the company that its product was not to blame.

The Prosecution claimed Martin-Baker should have *'designed-out any risk of human error'*. But this can cause a design to become so complex, so inherently unreliable, it becomes unsafe and increases the probability of system failures. An ejection seat is required to work once, and once only. It must be as simple as possible, avoiding the need for frequent or complicated servicing. One can error-proof a design to a degree, but defences in depth must still be erected, such as adequate training and its proper application. MoD failed to deliver these critical dependencies.

*

The Defence replied, citing MoD's own words:

> *'The first line of defence in Air Safety is the establishment of a questioning and learning culture, underpinned by the application of common sense'.*[127]

Had no-one questioned the decision to deviate from Martin-Baker's instructions, or learned from previous fatal errors? It was noted the

---

126 For example, Figure 21 in AP109B-0131-1. Also, AP109B-0131-5. Version held is at AL26, April 1992.
127 Regulatory Article 1020, paragraph 7.

company had been told to assume equipment would be properly serviced and installed. That, they did not accept or comprehend the Health and Safety Executive's insinuation that the 1949 design should have been computer-aided (impracticable until the late 1970s, with trial and error reduced by the advent of solid modelling in the 1980s).

The British Aerospace query of 1990 regarding *1.5 threads protruding* was raised. (See Chapter 12). The Prosecution misrepresented this as a long-term failure that caused Sean's death. Mr Tedd: *'There was a risk to many pilots over a lengthy period'.* The Judge later repeated this in a prejudicial sentencing remark:

> *'A significant number of pilots, and also potential passengers, were exposed to the risk of harm over a lengthy period'.*

Both ignored that the long-term risk was knowingly caused by MoD. One would expect Mr Tedd to repeat his client's claims, and to do so knowing he has been lied to is an ethical issue he must deal with. But for the Judge to accept something so plainly wrong was inexplicably inept.

*

But while interesting (in that it revealed the Prosecution's willingness to lie and pervert justice), the entire discussion was <u>unrelated to the causal sequence</u>. The red herrings were beginning to stink.

Day 2

Richard Matthews QC led for the Defence, conveying Martin-Baker's regret and, given the guilty plea, apologies for their *'contribution'* to the accident. (It would have been interesting had he been asked what that was). Having made the point others were involved, he asked: *'What were the standards expected of <u>them</u>?'* He could say little more, as the Judge had already stated MoD was not on trial. This was obviously a warning - if you want to reveal the truth, plead Not Guilty. But that does not mean she should have casually permitted the HSE's lies.

*

The *1.5 threads* claim was re-examined yet again. Tornado groundcrew at RAF Lossiemouth and Coningsby fitted the nuts to flush. This contradicted the Prosecution claim that *1.5 threads* was an absolute requirement, and disproved the lie that MoD never any information. The Prosecution claimed this was irrelevant as the Tornado had a Mk10A seat - a degree of compartmentalisation and incompetence that characterised its case. Mr Matthews reiterated that there were two risks. One of the nut coming undone, the other of over-tightening - both negated by fitting the nut and bolt properly. Correctly, he emphasised the only risk in Tornado was the former (because the Scissor Shackle is no longer there), but did not explain why this was exculpatory. Serving to emphasise MoD's failure to understand these basics, the Aircrew Manual was amended in May 2019 to instruct aircrew to check before flight that a minimum of 1.5 threads were visible. Yet, the accompanying illustrations show the head of the bolt flush.[128]

The evidence of a Chief Technician Morgan was cited - a new nut and bolt must be used each time. This is taught to all MoD apprentices and is set out in seat publications, but no-one seemed to appreciate it was yet another reason why the Routine Technical Instruction was rogue.[129] And again, someone knew, so the focus should have been on why others did not. No-one asked these witness how they knew. One always returns to the lack of consistent training and its application.

*

At the 2014 Inquest a witness had been Mr Martin Lowe, MoD's 'Subject Matter Expert' on ejection seats. He had also acted in this capacity at the 2010 Inquest into the death of Mr Mike Harland. (See Chapter 24). Now, the Defence quoted Mr Lowe's testimony confirming Martin-Baker servicing manuals were vetted by MoD. He had agreed the content assumed competency, expressing disappointment at the level of training available for Hawk. The Red Arrows had been described as *'not a good place to be'* for groundcrew.

The Judge asked why Mr Lowe was not present. (He was not the only one. Having conducted public lectures on the case, the former Hawk Type Airworthiness Authority might have been able to explain non-existent Safety Cases). Mr Matthews replied that Mr Lowe was a prosecution witness, and all he was doing was using his testimony to query the *1.5 threads* 'requirement'. That, it was the Prosecution who had introduced the differing standards applied at RAF stations. (The problem was, the Prosecution had implied Martin-Baker were to blame, and nobody bothered refuting this). Was Mr Lowe absent because the MoD/Prosecution did not want him exposed to questioning about why the accident was a recurrence, especially given the validity of the Safety

---

128 AP 101B-4401-15, Chapter 9.66.
129 For example, AP109B-0132-5F Bay Servicing Schedule, paragraph 89. Version held is at AL33, September 1993.

Case had been a key feature of the Mike Harland proceedings in 2010?

*

The Judge asked how a torque wrench worked, neglecting that the instruction was not to torque the nut. Also, if a competent engineer would be able to identify the over-tightening and pinching. This led the discussion away from the key point - that maintainers had been instructed not to conduct the check which would identify any pinching. The Defence replied that a competent engineer would not apply abnormal force. The Prosecution challenged this, but withdrew when invited to discuss it. Clearly it did not want to go there, because examination of the issue would reveal that it knew the wrong tools had been used, its claims were fallacious, and that its ally, MoD, was guilty.

Dr Simon Jones, an expert witness, had stated at the Inquest it would not have occurred to Martin-Baker that a trained technician would over-tighten the nut to the extent of cutting new thread (and bending the bolt). He confirmed it was only necessary to tighten the nut so it touched the shackle (i.e. the advice Martin-Baker provided). That, if the shackle assembly had been treated at all times as it was designed, the risk would not have occurred. The Judge remarked that Dr Jones had *'no aviation experience'* - perhaps unaware the subject was basic engineering fitting practice, regardless of trade. The Defence did not comment further, but should have been concerned at this dismissal of expert evidence due to complete misunderstanding of the subject.

The Defence emphasised the different maintenance policies of users, citing the US Navy practice of not carrying out *in situ* servicing. Also, that the US Navy used a Kynar (fluoro-thermoplastic) 'gripper' nut. The Judge accepted this and ruled that the McDonnell Douglas query *'had no relevance to Red Arrows, or RAF Hawks generally'*. That, the query *'did not put Martin-Baker on notice about anything to do with the RAF, who are supposed to be using common sense'*. We shall return to this shortly.

The Defence confirmed it was not blaming any individual, because MoD had admitted training was inadequate. It quoted a witness who had said it was *'unacceptable'* for an armourer to be posted to the Red Arrows having never worked on ejection seats.

Alluding to technician training, the Prosecution replied that inadequate training was immaterial; a nonsensical, asinine claim. This ignored that untrained superiors had disregarded airworthiness regulations and made false record. Nobody pointed out that adequate training is a prerequisite to a valid Master Airworthiness Reference.

Confused, Mrs Justice Carr expressed concern at being taken into the *'bits and bobs of an argument'*. This was not a retrial, and if the Defence wanted to go there, she may have to adjourn. Not for the first time she opened the door for Martin-Baker to withdraw their plea. But Mr Matthews was constrained by his client's position. I'm afraid Martin-Baker's reluctance to upset MoD and the US overshadows this entire case.

Hearing over, Mrs Justice Carr set sentencing for 23 February 2018.

Discussion

At the risk of stating the obvious, a feature of these proceedings was that the Defence, alone, spoke the truth. However, it refrained from using the whole truth, which did Martin-Baker no favours. One is entitled to ask why it failed in this duty, actively misleading the court.

The Health and Safety Executive was undoubtedly deceived by MoD, but proved itself incompetent, morally debased, and grimly determined to ignore the facts. It told the Judge it had undertaken a full review of the fresh evidence. This was a lie. It claimed *'many experts'* were involved, while refusing to heed or disclose expert evidence dismantling its case. It did not want witnesses exposed to probing questions revealing the extent of MoD's (and its own) failings.

This same scenario arose during the Chinook ZD576 case (Mull of Kintyre, 1994). Senior RAF officers had blamed the two deceased pilots, but MoD did not want them questioned at the Fatal Accident Inquiry so did not call them. MoD's QC then argued successfully that if the author of evidence was not present, it could not be cited. Perhaps this legal nicety was what the Judge was referring to when asking where Mr Lowe was? In both cases MoD was lauded for a thorough investigation, despite missing that the Master Airworthiness References were invalid. This failure was repeated in other fatal accidents, such as Nimrod XV230, Tornado ZG710, Hercules XV179, and Sea Kings ASaC Mk7 XV650 and XV704. (A total of 63 killed). Yet again, it had slain its own.

Mr Jim Cunningham

On the evening of 12 February 2018, I spoke to Sean Maffett. We agreed it might help if Mr Cunningham was told the truth. He wasn't going to get it anywhere else. A short letter was prepared.

At lunchtime the following day *Mr Cunningham approached Sean*, who passed him the letter. (Sean had given our details to Mr Cunningham's QC, Tom Kark, on 23 January 2014 during the Inquest, so he at least recognised a friendly face). It was snatched from his hand by MoD 'minders'. After the day's hearing, Mr Cunningham again approached Sean. Asked *'Did you read it?'*, he replied *'No, they took it away from me'*. He had been told it was a leaflet being circulated by journalists. Before being bundled into a car, he managed to say he was *'horrified'* by the decision to prosecute Martin-Baker. Another minder stepped in: *'He's said he doesn't want to talk to you'*.

This behaviour was nothing short of outrageous. It reminded us of a similar incident in 2008 at the Hercules XV179 Inquest in Trowbridge. Retired RAF pilot Nigel Gilbert, who was to be a witness, was detained on the courtroom steps by MoD Security Forces. Luckily, I had already provided evidence to the Coroner's Investigator and a family's QC, revealing MoD's lies. This took the form of two MoD specifications for Explosion Suppressant Foam, dating to the early 1970s, disproving its claim never to have heard of the foam until after the accident in 2005. Like here, this revealed that a known risk was not ALARP. The difference was timing. The XV179 Board of Inquiry report was released before the Inquest, so could be assessed early enough for the public and Coroner to take action.

What Sean found 'interesting' was the entire Prosecution team knew who he was. As he had dealt exclusively with the court, in his capacity as an accredited journalist, plainly a third-party had briefed the Prosecution. One only has to ask who benefited.

Another development was Sean receiving a number of calls on 13 February from a family friend of the Cunninghams, a former Red Arrow pilot who had known Sean Cunningham well. He wanted to expand on what we already knew from the Service Inquiry - the standard of engineering at the Red Arrows was poor due to cutbacks forcing dangerous shortcuts. Also, that the RAF had tried to court martial junior staff while ignoring directed violations by senior officers.

MoD and the BBC

I mentioned the BBC had promised to get back. On 21 February 2018 we were provided with the text of an e-mail from MoD:

'I am pleased to report that the facts of this matter pretty much match that of our previous discussion:

The type of shackle fitted to the seat was and is not the issue here, both types (scissor and gas) are adequate for their purpose, the issue as identified by the Coroner, was that <u>specific guidance</u> on the degree to which the nut was to be tightened on the scissor shackle, and the potential consequences of over-tightening, were not issued to the MOD/RAF by Martin Baker.

Several variants of the Martin Baker Mk10 ejection seat are in service with the RAF. The Hawk T.1 aircraft is fitted with the Mk10B seat. The Mk10B is a compact, lightweight seat suited to smaller aircraft such as the Hawk. The Tornado GR4 has a different seat fitted (the Mk10A) which is larger and heavier. The modification programme for the Tornado was to deliver 0-0 performance for a 130kg boarding mass occupant, (=large man in full combat survival gear) this was achieved by installing a much larger <u>GQ5000</u> parachute system into the seat as opposed to GQ1000 on the Hawk, it was this fitment that led to the introduction of the gas shackle.

In sum, replacement of the scissor shackle was/is not required on the Hawk T.1 as it <u>remains fit for purpose</u>, the gas shackle was fitted to the Tornado due to the increased weight requirement and enlarged parachute.

Grateful for confirmation that the above info is understood and that this line of enquiry is now dropped'.

We thought the final sentence an outright threat, which is routine. At the trial in January, an MoD media relations official had accosted a young BBC reporter demanding she hand over her notes (from myself and James Jones). To her eternal credit, she refused; and probably explains her supplying this e-mail. James and I were asked for comment. We replied:

- Martin-Baker <u>did</u> issue *'specific guidance'*. MoD stopped using it.
- The Gas Shackle had nothing to do with the adoption of the GQ5000 parachute, confirmed by the QinetiQ report of 2002.
- Implicit in the statement that the Scissor Shackle remained in Hawk as it was *'fit for purpose'*, was that MoD understood how it worked and how it was to be serviced.

MoD did not want the Gas Shackle aspect explored publicly, for two reasons. First, it had denied all knowledge of the issue to a Minister in 2014, who had then misled Parliament. (See Chapter 13). Second, and demonstrably, Martin-Baker were acutely aware of failure modes and effects, and dealt with them correctly - by design, and by issuing comprehensive servicing instructions. Both MoD and the Health and Safety Executive would be concerned at the truth being broadcast.

The BBC was asked to put our points to MoD. While the reporters on the ground recognised the issues, and had gone to the length of conducting an on-camera interview with James on 30 January 2018, the final brush-off came on the evening of 21 February:

'After a lot of discussion, thought and also speaking to other people with aviation expertise, we feel we cannot include the concerns you have raised in a short news piece on the day of the sentencing of Martin Baker. That's not to say we can't continue to look at this, but we feel the time and level of detail needed to explain your concerns is much greater than our allotted slot during Friday's broadcast'.

It declined to say who these anonymous experts were, or what subject they held forth on, but clearly the BBC had succumbed. Not only that, it continued reporting allegations it now knew to be untrue, negating the possibility of it later explaining our concerns.[130] And it did not.

The Times

Our dealings with Defence Editor, Deborah Haynes, followed the same pattern. Numerous requests for details, accompanied by promises to print a major piece. But her legal department refused to allow it to run unless we could *'prove the Health and Safety Executive erred'* - proof already provided in MoD's hand. When we supplied it again, there was no reply. Separately, the News Editor asked: *'Where is the proof the shackle was screwed on too tight?'*. You can lead a horse to water...

Day 3 - Final sentencing hearing

At 1008 on 23 February 2018 the Red Arrows conducted a fly-past over the courthouse. Whoever authorised this cunning stunt should be held in contempt. A little humility would have been in order, given the actions of their superiors that killed their colleague.

Before entering court, Mr Cunningham again approached Sean Maffett, asking that they meet later. He was again hustled inside by a 'minder'. This was witnessed by the BBC's Gemma Dawson.

---

130 https://www.bbc.co.uk/news/uk-england-lincolnshire-43171049

Mrs Justice Carr had received a Victim Impact Statement from the family the previous evening. The Defence noted that it referred to issues which had not been before the Judge, and they sought to discuss it with her to ensure her Remarks would not contradict it. The Prosecution had read the Statement, and had no comments.

The Judge then spent some time summarising the case. She stated that *'on or before August 2011'* Martin-Baker had failed to ensure that persons not in their employment were not affected by parachute failure. Thus, she confirmed the timescale being looked at was no longer 1990-92, but any time before August 2011 - in effect, accepting the HSE's claim, but without requiring the formal charge to be amended. In saying this she overlooked all the evidence she had been given going back to 1952. This set the tone for the day's proceedings, and her formal Remarks.

She confirmed risk was an ingredient of the offence, but the actual harm was not. That, there was *'absolute criminal liability'*, and *'good systems in place, but they were not sufficiently adhered to'*. She was referring to Martin-Baker, but knew that it had been MoD who flouted its own system.

She stated that in 1990-92 Martin-Baker were aware of the risk, but had not informed MoD. She knew this was wrong; that the company had mitigated the risk by 1984, and MoD rejected the mitigation. So by definition had been informed.

Similarly, she repeated the evidence of Mr Butter, who had claimed the Scissor Shackle arrangement should have been redesigned to avoid the risk. Again, this ignored that Martin-Baker had redesigned the release mechanism, by introducing the Gas Shackle.

She said that the *'express requirement was to provide safety in all situations of flight'*. Again, ignoring that it was MoD who systematically refused to meet this legal obligation, and that it had killed scores.

She concluded that Martin-Baker were culpable, and had a duty to warn RAF engineers about over-tightening. She omitted that many such warnings had been provided since the 1950s, but that MoD chose to ignore them. And, that she had even named an MoD recipient. (This bears repeating - it is the most bizarre and farcical aspect of the entire proceedings).

Mrs Justice Carr did, however, acknowledge MoD's admission that training within the Red Arrows was poor, and that concerns had been expressed long before the accident. She also stated that Martin-Baker's argument that common sense dictates one should not over-tighten the nut was valid.

The remainder of the day was devoted to arguments over degree of liability and Martin-Baker's financial status, which would determine the level of fine. They were fined £1.1M for not providing information they had already provided, but which maintainers were instructed not to use. Applying this information would have saved Sean's life. Two days later the Judge's Remarks were published, which I discuss in the next chapter.

Speaking to the family, she said: *'I pay tribute to the courage and dignity of the family throughout'*.

In a statement, Mr Jim Cunningham said:

> *'I would like to address the MoD and RAF. I have one question for you. Where are you? Not one of you is here to represent my son or my family. Sean was so proud to be a member of the armed forces and to serve his country. He was the perfect ambassador for your organisation. He gave you his all and ultimately gave his life'.*

I'm sorry Mr Cunningham. MoD lost interest in Sean the moment he died. Its only concern was concealing the truth from you.

21: Sentencing remarks of the Honourable Mrs Justice Carr, DBE, 25 February 2018

'It is a fundamental principle of judicial conduct that one cannot disregard what the evidence establishes'.

Judge Philip Boyd, 14 March 2013

Mrs Justice Carr was Senior Presiding Judge for the Midlands Circuit, previously specialising in professional negligence and product liability. As such, she had both great power and great responsibilities. She would know this was a high-profile trial with potentially far-reaching consequences. The truth would open Pandora's Box, exposing misconduct by the same Crown Servants who had been unmasked in previous cases, but had escaped prosecution.

Martin-Baker's decision to enter a late plea of guilty undoubtedly confounded her, as the evidence had already established MoD's culpability. The basis of the Prosecution's case - the two queries - had been systematically demolished by her acceptance of the Defence evidence, and reinforced by the Prosecution's inability to even comment. Nevertheless, she allowed the case to continue. Onlookers were thoroughly bewildered by this quirk of the legal 'justice' system.

She now had to decide what to say in her Sentencing Remarks. On 13 February 2018 she called this a *'most unusual sentencing event'* and did not *'know how to continue'*. Most took this as a tacit admission of puzzlement over the guilty plea. One wonders, then, to whom did she turn for guidance? Justice Haddon-Cave? He would have told her the wrong party was in the dock.

Like Coroner Stuart Fisher before her, Mrs Justice Carr was seriously misled. Martin-Baker's plea placed her in an invidious position. Mindful of her need to construct her Remarks to fit the plea rather than the facts, let us examine what she said, and compare it with the truth. For brevity, legal discussions and digressive comments are omitted. The complete remarks are available to download.[131]

---

131 https://www.judiciary.gov.uk/wp-content/uploads/2018/02/sentencing-remarks-mrs-j-carr-r-v-martin-baker-lincoln-crown-court-23022018.pdf

R (on the prosecution of Her Majesty's Inspectors of Health and Safety)

v

MARTIN BAKER AIRCRAFT LIMITED

Introduction

Martin Baker Aircraft Limited [MBAL] now stands convicted on its guilty plea entered through one of its statutory directors, James Martin, on a single offence of failing to ensure the safety of non-employees contrary to s. 3(1) and 33(1)(a) of the Health and Safety at Work Act 1974 [the Act]. The particulars admitted are that MBAL, on and before 8 November 2011, failed to conduct its undertaking in such a way as to ensure, so far as was reasonably practicable, that persons not in its employment who may have been affected thereby, including Flight Lieutenant Sean Cunningham, were not thereby exposed to risks to their health or safety, in particular by non-deployment of the main parachute attached to a Mark 10B ejection seat during low speed or zero-zero ejections.

Indirectly, this confirmed the charge related to the main parachute not deploying, and encompassed the period 1952-2011. However, she had apparently changed her mind about the end date, moving it from August to November 2011.

It is convenient to record here that, due to the passage of time, not all of the relevant papers are now available.

Did this include the papers provided to her by the public and Coroner which MoD and the Prosecution had not disclosed?

The RAF/MoD is neither a party nor represented in these proceedings.

Strictly speaking true, but Lieutenant Commander Graham Hamilton RN, a contributor to the Service Inquiry report, was in court acting for the Prosecution.

The submissions before me have nevertheless descended into considerable detail. I received over 100 pages of submissions and two full bundles of dense, often technical, documentation. But I emphasise again that my function is not to conduct a trial of the merits. I could not in any event do so without areas of disagreement being explored fully on the evidence. I expressed concerns during the sentencing hearing about the appropriateness of being taken to only excerpts from statements or reports, without understanding their full context or being told the extent to which their contents were agreed or not agreed.

'Descended' seems disparaging, tending to confirm she did not want to hear any detail, regardless of how germane it was. And if she did not understand dense technical documents, she had the wherewithal to take specialist advice - which, after all, would have been succinct. And while there may have been *'areas of disagreement'*, there could be no doubting the most basic fact - the Health and Safety Executive accused Martin-Baker of never providing information that it knew MoD had always possessed, and the Judge had named a recipient.

> *I have received, either directly or through my clerk, a considerable amount of unsolicited correspondence and material from third-parties in relation to this matter. In order to ensure the integrity of the process, I have not read this material but rather passed it on to the parties for them to consider what, if any of it, is necessary or appropriate for me to consider. I have not in fact been invited by them to include any of the material in my deliberations.*

She then contradicted herself:

> *The Martin-Baker ejection seat has been an important and valuable air safety development, designed for use in an emergency, and which has undoubtedly saved many lives over the years. So much is clear from the testimonials and letters I have read from pilots who have ejected successfully in the past.*

Addressing the design of the seat:

> *Since about 1984 MBAL has not designed any new seats with a scissor shackle. Rather it uses an improved gas-release shackle system, available for new aircraft and retro-fitting. The MoD contracted MBAL to carry out such retro-fitting on all in-service ejection seats, with the exception of the seats in the Hawk aircraft.*

This confirmed MoD knew of the risk of shackle pinching and how to eliminate it.

8 November 2011

> *On Friday 21 October 2011, Hawk XX177 was due for a routine inspection to ensure that there was no cracking to part of the ejection seat block assembly. Two RAF groundcrew carried out the work that day. Following the inspection, and non-destructive testing, the shackles were reconnected on Monday, 24 October 2011. During the reconnection, one of the technicians tightened the nut onto the bolt of the drogue shackle to 1.5 threads. There was no instruction to the engineer to the contrary.*

The *'routine inspection'* instructions were unsafe.

There was no requirement for an instruction regarding thread count, as the concept did not apply in this application. There *was* an instruction on how to service seats and ensure serviceability.

Whilst it was possible, to a limited degree, to check whether there was free movement between the shackles, it was not possible to check whether the scissor shackle could be released.

But the maintainer must not have checked for free movement. The degree of over-tightening meant the shackles were clamped together, witness the rub marks on the Scissor Shackle. (Figure 9).

Subsequent investigations revealed that after the nut had been tightened onto the bolt on 24 October 2011, the width of the scissor shackle was wider than the gap between the outer ends of the lugs on the drogue shackle, leading to an interference fit impeding the shackles' separation. The lugs were not parallel, having tapered. The drogue shackle had been tightened with sufficient force to bend the drogue shackle bolt and cut new thread on the bolt. The interference fit would not have been detectable by the groundcrew. Later testing, led by Lt Cdr Hamilton, demonstrated that the current method of installing and tightening the nut and bolt on the drogue shackle introduced a hazard that could prevent its release, resulting in failure of the main parachute to deploy.

The *'current method'* (using the wrong tools and techniques) and the associated hazard were introduced by MoD.

Following the incident, the RAF/MoD requested that a shoulder bolt be used. The same modification has been offered by MBAL to, but declined by all other organisations around the world still using the same mechanism.

Most inferred this was MoD insisting on first-time use of a shouldered bolt. In fact, it had previously agreed to its removal from the Mk9 seat. There are many potentially valid engineering reasons, but to explore this area would cloud the main issue. That, irrespective of bolt type, no serviceability check was conducted, but the seat was declared serviceable.

Culpability

This admitted breach arises out of the fact that in early [probably February] 1990 Mr Alan Lowther of MBAL wrote a note [on a MBAL compliments slip] to Mr Mackie in the Quality Assurance department as follows: "...F18 Info as requested. We are going to put similar info into our pubs, i.e. similar illustration to attached and necessary instructions but no dimension for clearance, only that there should be clearance and scissor shackle should not be pinched. NO TORQUE LOADING".

As I noted in the previous chapter, the Judge had already ruled that the

McDonnell Douglas F18 query *'had no relevance to Red Arrows, or RAF Hawks generally'*. That, the query *'did not put Martin-Baker on notice about anything to do with the RAF, who are supposed to be using common sense'*. Yet the charge only related to the McDonnell Douglas query, which the Judge had ruled had *'no relevance'* to the Red Arrows, or to Martin-Baker's obligations to MoD. <u>In other words, Martin-Baker admitted to something the Judge then ruled was not a breach</u>. On the face of it, this is a serious anomaly and needs resolving. It may be poor phrasing. But it may be an error which, if Martin-Baker were minded, would be grounds for appeal. There was no jury to misdirect, but did the Judge misdirect herself?

Setting this issue aside, Mrs Justice Carr was obliged to accept the 'judicial truth' of Martin-Baker's guilty plea, despite knowing it was contradicted by the real truth. Nevertheless, the jottings on a compliments slip are not the official company position. Using them out of context in this way was prejudicial.

In any case, only MoD can decide what to do with its publications. It already had the necessary information and more, but had decided to no longer use it.

> *Thus, by this time at the latest, MBAL was aware that it needed to issue a warning in its publications relating to clearance and warning against pinching of the scissor shackle. There should have been a warning to guard technicians against over-tightening the drogue shackle locknut. For reasons which are not explained, this never happened, at least so far as the MoD/RAF were concerned.*

Notwithstanding the previous issue, this compounded her error. The compliments slip jottings would have to be assessed from a legal perspective before being issued by the company. This process can be seen in the Special Information Leaflets of 2011/13, where Martin-Baker reiterated that *'<u>All</u> safety and maintenance notes detailed in all relevant technical publications are to be complied with'*. That is, check serviceability.

> *By this breach MBAL exposed each RAF pilot [and any passenger] flying a Hawk to a material risk, namely that if the pilot was ejected from a Hawk in zero-zero or low speed conditions, the two shackles might not release from one another.*

> *The Prosecution case, however, goes beyond this admitted breach. It criticises MBAL's design from inception...in 1971...*

The breach was by MoD. Moreover, and importantly, that makes every person who has flown in a Hawk an 'interested party'; meaning she should have read and ensured she understood their evidence. She did not.

Another factor this introduces is that the Red Arrows often fly civilians as a passengers - in fact Sean Maffett, whom I have mentioned, flew with them in his capacity as a BBC journalist. He presented himself as a witness, but was not treated as an interested party.

Shackle design *'inception'* (In Service Date) was 1952. Using 1971 ignored 19 years of prior notifications by Martin-Baker, and was prejudicial.

The Prosecution contends that MBAL failed to produce an assembly drawing showing the components fitted together, only separate drawings for each component.

The Defence throws credible doubt on the suggestion that MBAL ought to have produced an assembly drawing. Each individual element was designed taking into account adverse tolerances, in the absence of a 'CAD' computerised design system.

Every servicing manual and MoD publication includes a general view showing the relative location and orientation of the release mechanism components; including the Barometric Time Release Unit, which is key to understanding operation. If (for example) cancellation of training in 1983 rendered existing instructions unclear in any way, that was for MoD to resolve by contracting amendments to the drawing set.

Thus the risk of an interference fit was not identified and the effect on the shackle dimensions once the nut and bolt were fitted was not shown.

The risk was identified. MoD chose not to eliminate it, either by checking shackle disengagement or introducing the Gas Shackle. The dimensions (Inner Lug Gap) remained correct if the nut and bolt were fitted in accordance with Martin-Baker instructions.

The Prosecution contends that MBAL never appreciated the recipe for disaster if there was standard tightening to 1.5 threads. This was an obvious risk, not dependent on the application of hindsight. It relies on the expert opinions of Mr Butter, HM Inspector, and Mr Rudland, a health and safety specialist inspector in the field of mechanical engineering, who state that the design of the drogue/scissor shackle assembly was poor.

Martin-Baker did appreciate the risk and advised MoD - witness their manuals, training, and adoption of the Gas Shackle in Tornado.

Tightening to *1.5 threads* is not *'standard'* on such a device. The

Prosecution argument was specious.

MoD disagreed with Messrs Butter and Rudland. It remains content with a design that has always operated correctly if Martin-Baker's instructions are followed.

It was reasonably practicable to design out any risk of an interference fit.

And Martin-Baker did so, by 1984 at the latest, but MoD did not adopt the design until 2007 in Tornado, and not at all in Hawk.

There is evidence that [since at least 2007] engineers at RAF Lossiemouth and RAF Coningsby have tightened the threaded fasteners used with the drogue shackle to flush. This experience undermines the Prosecution's case that the standard of 1.5 threads is applied universally in the absence of instruction to the contrary.

The Judge sets out the best case for the Defence. That is, the information MoD said it never had was at other Air Stations. This suggests she was led to believe Martin-Baker were responsible for not disseminating it correctly, and ensuring it was implemented - which was the Prosecution's stated position.

The design and manufacture were founded on the premise that the specification requirements applied to equipment which had been correctly installed and serviced [as evidenced in contractual correspondence in 1972]. It appears to be recognised that an engineer is expected to use his/her training to interpret instructions in Air Publications. As Mr Lowe, the subject matter expert for the ejection seat, has commented, MoD Air Publications assume a level of engineering competency.

This challenges MoD's claim that groundcrew have Defence Standard 00-970 to refer to. As the seat had not been correctly serviced - by conscious policy, not omission - in effect the Judge again pronounces innocence.

There is evidence that 'abnormal' force was applied at the time. New thread had been cut on the plain shank of the bolt close to the unthreaded portion. The presence of two scores indicated that a nut had either been fitted twice or backed off and re-tightened, with the faces of the bolt's threads undergoing adhesive wear.

In such a discussion one should note the requirement to (a) always fit a new nut and bolt, (b) not torque the nut, and (c) ensure disengagement.

There is no mention of the degree of *'abnormal force'* (see Figures 8 and 10), or that the RAF chose not to use the tools provided by Martin-Baker.

Both would be highly detrimental to the Prosecution's case.

MBAL raises issues relating to the quality of the training of RAF/MoD engineers on the Hawk at RAFAT after 1983. Reference is made to various statements, in particular one from Martin Lowe, head of engineering for aircraft assisted escape systems for Defence Equipment and Support, a branch of the MoD, who describes the lack of formalised or structured training packages. Reference is made to a MoD quality audit report from January 2011 referring to RAFAT technicians being 'left out on a limb'.

Again, the uninitiated would infer this was a one-off, applicable only within the Red Arrows. From 1992-on the RAF Director of Flight Safety had reported this was a systemic failure.

The Prosecution points to the fact that any failures in training did not contribute to the incident on 8 November 2011.

This was not a *'fact'*. It was a lie.

As for the correspondence with McDonnell Douglas and BAe [British Aerospace], MBAL states that it did not relate to the risk of interference fit but, at worst, to the risk of a 'hang-up' [a recognised risk of a momentary delay in the release of the mechanical drogue shackle as it aligns during the ejection process]. The McDonnell Douglas correspondence, rather than relating to the inherent risk of clamping as a result of the drogue shackle's design, requested MBAL's concurrence to the US Navy's removal of its own requirement when servicing F-18 seats to torque load the locknut to 60-85 inch-pounds because it was not meeting its own requirement for 0.030" clearance. The US Navy does not carry out on-plane servicing.

This discussion of a January 1990 query omitted that other US servicing manuals, from 1974, said *'Do Not Torque'* - placing the query in an entirely different light.[132] The existence of this 1974 instruction cannot be reconciled with the Judge's remarks.

The Judge omitted that this possibility of *'momentary delay'* was fully explained in the 2002 QinetiQ report, undisclosed by MoD but provided to both her and the Prosecution by the public.[133]

It is unclear why the Judge is discussing the McDonnell Douglas query,

---

132 United States Department of Defense Technical Manual TM 55-1680-308-24. Copy held is dated 13 December 1974.
133 QinetiQ report AT&E/CR00782/1 'Mk10A Ejection Seat Modifications (02097 & 02198) for Tornado GR4/4A and F3 Aircraft - Phase 2', paragraphs 3.1.4 and 3.1.5, December 2002.

given she has already confirmed it is irrelevant to the Red Arrows.

Ultimately, resolution of the debate would not advance matters materially for my sentencing purposes, even had I been able to resolve it on the criminal standard of proof [which I have not been].

First, as MBAL itself points out, it did not need to be put on notice of the risk of over-tightening. That was a matter of common sense.

Secondly, at around the very same time - early 1990 - MBAL in any event identified the need for better warnings. This is the basis of its guilty plea. In this sense, it is common ground that it was on notice by this stage of the need for more action, which it failed to implement. That breach should be seen in broad context, including that it was committed in circumstances where MBAL was nevertheless entitled to expect the application of common sense by properly trained RAFAT engineers when fitting and servicing the shackle assembly in question.

'Better warnings' is a red herring. That does not mean the existing warnings were inadequate. MoD had declared itself fully content with them, given the training it provided to its maintainers at the time. By ceasing this training, <u>MoD</u> created the need for a better warning. The key question is what did MoD ask of Martin-Baker? This was not asked. For example, MoD would need to update its Training Needs Analysis, so Martin-Baker could pitch the instructions at the correct level.

The US Navy inserted a torque figure in its F/A-18 manuals, when Martin-Baker's explicit instruction was not to. The indications are that this was an isolated error, as other US manuals said *'Do Not Torque'*. This point is crucial. The query did not apply to MoD which, like the US, simply needed to heed existing warnings. And there is only one way of saying one must certify serviceability before signing to say an item *is* serviceable. Martin-Baker had also provided that warning, reflected in MoD publications.

The Prosecution concentrated on shackle reassembly while ignoring the need to check serviceability, when they are part of the same maintenance operation. Hence, the Judge (and the case) <u>proceeded on the wrong basis</u>.

Offence category: culpability and harm

There are features of high culpability offending including breach over a long period of time. The breach persisted from the 1990s onwards through to November 2011. It was not an 'isolated' breach in this sense.

The Judge contradicts the *'on and before 8 November 2011'* timeframe in her opening statement, now limiting it to '1990s-onward'. In doing so she rejects or ignores exculpatory evidence from the previous 38 years.

The Air Ministry knew of the risk of over-tightening, the effect and criticality, in the 1950s, as did MoD in 1964. It was repeated in 1984 (Gas Shackle), 2002 (QinetiQ report) and 2007-10 (embodiment of Gas Shackle). The Air Ministry took mitigating action. MoD terminated that action, committing conscious and persistent breaches in the face of uncompromising warnings from the RAF Director of Flight Safety.

The law is not a moral compass

The Prosecution's case relied entirely on speculation, which neither it nor MoD could prove. Nor did they provide any evidence even remotely suggesting guilt on the part of Martin-Baker, who were judged against the wrong standards.

When the Mk10 seat was introduced, the RAF had the correct information to implement its Maintenance Policy, and groundcrew were trained accordingly. MoD freely admitted it changed the Maintenance Policy, ceased training armourers, and decided to ignore Martin-Baker. At that point, it was incumbent upon MoD to make compensatory provision; for example, by contracting all seat servicing to Martin-Baker. Recognising this, the company submitted a bid, rejected by MoD. That bid is exculpatory evidence.

But Mrs Justice Carr's role was to adjudicate, not investigate or go beyond evidence led by the Prosecution or Defence. She had to accept the guilty plea and abide by rules - rules which are a simplified product of our morals. But they can never cover every eventuality, which is why we must look for the application of professional judgment and good old common sense. Natural justice demanded instant dismissal of the Prosecution's case, and a recommendation that the Crown Prosecution Service reconsider its position regarding MoD.

In addition to rejecting key MoD and Prosecution claims, Mrs Justice Carr mentioned eight reasons why Martin-Baker were not culpable:

- The hazard was introduced by MoD.
- Maintenance errors during reassembly.
- No disengagement check, thus failing to ensure serviceability.
- Information known by other maintainers.

- Poor training.
- Previous warnings.
- The Gas Shackle.
- The admitted breach (which she confirmed did not take place) was irrelevant to the accident.

To make these statements, she must have been satisfied the evidence was properly led, and accurate. However, she was misled into misinterpreting large parts of it, and guided or affected by extraneous matters. Unaware of the boundaries of responsibility between and within MoD, the RAF, Martin-Baker and British Aerospace, she confused their roles and responsibilities. But even if she did not grasp the full details, alarm bells should have sounded when MoD documents disproving the Prosecution case were produced by a member of the public. And especially when Coroner Stuart Fisher forwarded the same evidence to her, and she drew heavily on his Inquest proceedings.

*

The Inquest had heard from one carefully selected witness who insisted he had never been told not to over-tighten the drogue nut. But we never heard from those who decided he should not be trained. Or those who changed the seat maintenance policy without updating publications. Or who issued the policy to waste money, prompting dangerous shortcuts to maintain some semblance of operational capability. Or who signed off risks as tolerable and ALARP, knowing they were not. Quickly, we get back to the same coterie whose policies and reckless indifference to legal obligations begat many other avoidable fatal accidents.

With the blame now successfully transferred, the likelihood of MoD continuing to offend is high. And it has, witness the grounding and subsequent scrapping of Air Cadet glider fleets, as a consequence of the same systemic airworthiness failings, overseen by the same people and the same regulatory authority. And the death of Corporal Jonathan Bayliss, a mere month after the Judge issued her Remarks. More aircrew will die unnecessarily. Who do you think will have contributed most?

PART 5 - FULL CIRCLE

The elements of misconduct in public office are: (1) a public officer acting as such; (2) wilfully neglects to perform his duty and/or wilfully misconducts himself; (3) to such a degree as to amount to an abuse of the public's trust in the office holder; (4) without reasonable excuse or justification.

Attorney General's Reference, #3 of 2003

22: Video evidence

'It is dangerous to be right in matters on which the established authorities are wrong'.

Voltaire

To recap, it was alleged that at no time between 1952 and 2011 did Martin-Baker provide information to MoD (or its predecessors) setting out the risk of the main parachute not deploying as a result of over-tightening the Drogue Shackle nut. The obvious line of defence was to provide proof that it <u>was</u> provided, and <u>when</u>; leaving MoD to explain why it was not used. The Judge and Health and Safety Executive were notified of the following <u>pre-1990</u> evidence by properly interested parties, refuting the allegation:

- Martin-Baker and MoD publications (1955-on)
- Aircrew and engineers' accounts of training (1958-on)
- The advent of the Gas Shackle (1984)

The Health and Safety Executive claimed none of this suggested MoD knew of the risk of shackle pinching before 1990.[134]

Video evidence

Incredibly, this proof of prior knowledge is outdone by a series of twelve <u>RAF</u> training films entitled 'Ejection Seat Servicing', released in January 1959.[135] Two sequences, of a Mk4 seat, show the instructor demonstrating the shackles must disengage when the scissor jaws are open, ensuring the release mechanism works as intended.[136] This design concept and functionality remains the same for the Mk10B seat.

The Health and Safety Executive denies this, saying the films are unrelated to the Mk10B design arrangement, so would not be subject to the same functional checks, and would not require the same

---

[134] E-mail from Health and Safety Executive principal investigator, 26 February 2018 11:43.
[135] RAF training films AMY 516/1-12 'Ejection Seat Servicing', January 1959.
[136] Video sequences at 61m 40s and 78m 10s. Available to view at https://sites.google.com/site/militaryairworthiness/red-5-2019

information.[137]

Figure 12 depicts the Scissor/Drogue Shackle and Barometric Time Release Unit Restraining Plunger arrangement in a Mk4 seat. Notwithstanding the Health and Safety Executive's claim, is can be readily seen that functionally it is identical to the Hawk Mk10B.

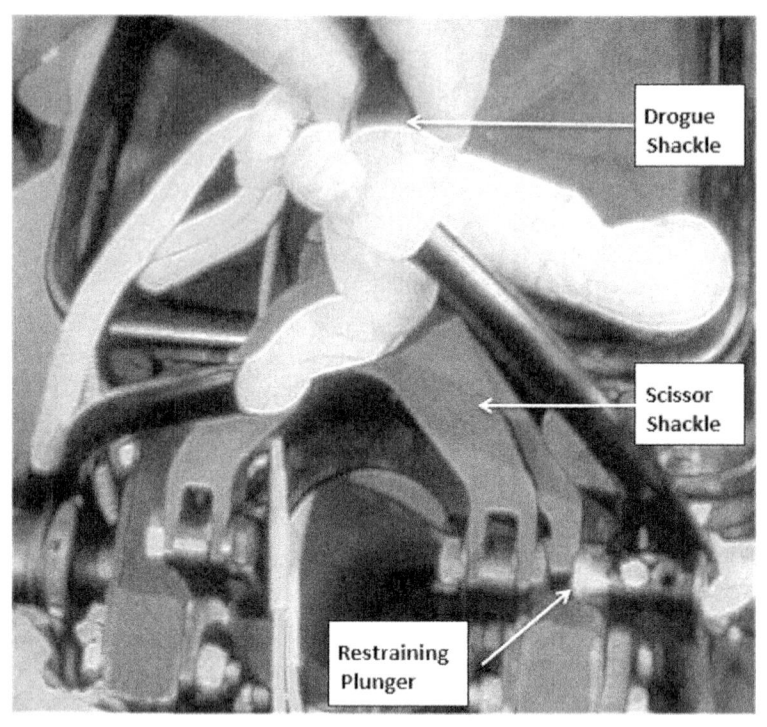

Figure 12 - Still from 1959 RAF training film, showing shackle disengagement check. *(Crown Copyright - expired)*

Having described the assembly process, the accompanying narrative says: *'Check <u>again</u> the release of the Scissor Shackle'*. Martin-Baker servicing manuals and MoD Air Publications set out where this is to be carried out - in the servicing bay, prior to cocking the Barostatic Time Release Unit.[138] Once completed, the mechanism must not be disturbed without

---

137 E-mail to author from HSE principal investigator, 17 May 2018 12:36.
138 AP109B-0131-5F Bay Servicing Schedule, paragraphs 4.6 and 41 NB2. And Martin-Baker Instruction and Servicing manual M-B/17, paragraphs 107b and 107g.

repeating the check. The Health and Safety Executive's claim that MoD did not have this information is again proven false.

The film ends:

> *'An ejection seat is only used in an emergency, so service it carefully. Pay great attention to detail and the pilot will be safe. Safety depends on YOU'.*

Precisely, so why did the RAF ignore its own training, and that provided by Martin-Baker?

A picture speaks a thousand words. Video screams. A simple internet search uncovered these films, and there are more. There were other clues. From Martin-Baker's 'School for Safety', discussing its training school established in 1955:

> *'The (school) is well equipped with training aids, including sectioned assemblies and films. One of the most important series of lectures is the official RAF maintenance course. The course lasts one week, with a whole day devoted to practical work, the majority of which is spent in stripping and reassembling time releases, drogue guns and other timing mechanisms'.*

MoD claimed to have no knowledge of what this training entailed.

Perpetuating and compounding the lie

The Health and Safety Executive's position was briefed to various trade magazines. For example, the February 2018 edition of IOSH Magazine (Institution of Occupational Safety and Health):

> *'The parachute did not deploy due to a mechanical fault Martin-Baker had known about since the 1990s'.*

The fault condition was created by MoD when applying the Routine Technical Instruction. The article also repeated the lie that MoD was unaware of the risk of over-tightening.

After sentencing, the Health and Safety Executive's Operations Manager, Mr Harvey Wild, said: *'This was an extremely complex investigation'*. No it wasn't. MoD admitted serious offences. The background had already been published by members of the public and the RAF Director of Flight Safety, and reiterated by Mr Charles Haddon-Cave QC and Lord Alexander Philip. It had been accepted in full by the Government. The only complexity was having to sustain your false narrative.

23: Follow-up

The Times

Defence Editor Deborah Haynes had declared herself unhappy over the court, and hence her newspaper, being misled, and had promised to run a piece explaining the truth. Despite the demands of her editorial staff (outlined earlier), on 25 February she asked me for a critique of the Judge's remarks. I sent her what was, effectively, an early draft of Chapter 21. I concluded:

> 'There has been no independent input to explain the MoD procedures which, if implemented, would have prevented the accident'.

The same day she put it to Mr Cunningham that he might wish to speak to us or the Times, perhaps both. He was *'very happy'* about this proposal. (Bearing in mind he had been actively prevented from speaking to Sean Maffett during the hearings). Deborah was equally enthusiastic:

> 'I agree with you on the curious timeline and the real reason behind the Martin-Baker's volte face. It would be powerful from a story telling perspective if Jim Cunningham is willing to meet with us. I would love the chance to observe as you sat with him and talked in the way you have shared your expertise with me'.

She was then diverted to the Skripal poisoning, returning on 17 April and intent on pursuing the story. But shortly afterwards she left the Times for Sky News. Her successor was *'keen to investigate'*, but I'm afraid we heard nothing more.

The Health and Safety Executive (HSE)

On 26 February 2018, the HSE's David Butter wrote to me:

> 'It is specifically denied the court was in any way misled'.

A practice is misleading if information is ambiguous or important information omitted. Lying by omission occurs when an important fact is left out to foster a misconception, including the failure to correct pre-existing misconceptions. I replied:

> 'MoD claimed it had no knowledge of the need to avoid over-tightening the nut. Both you and MoD knew this to be wrong, and held the evidence that it was wrong'.

Mr Butter replied claiming this was *'irrelevant'*.

*

On 23 March 2018 I sent an e-mail to Mr Butter, copied to Mrs Justice Carr, Lincolnshire Police and Irwin Mitchell Solicitors (who had been representing the Cunningham family):

> *'The purpose of this e-mail is to advise you that evidence in the form of twelve RAF training films has emerged proving maintainers were taught, before 1990, to ensure the Scissor Shackle and Drogue Shackle of Martin-Baker ejection seats must be able to freely disengage when the Scissor Jaws are opened.*
>
> *There is now proof the RAF knew the ultimate test was the ability to disengage. It was, and remains, MoD's responsibility to use that information properly.*
>
> *[I attached a still from the video - Figure 12]. I believe this is exculpatory evidence. You will appreciate the operator is holding the Drogue Shackle in his left hand, and checking disengagement from the Scissor Shackle. Had this simple operation been carried out, the failure to disengage would be immediately evident, it would (one assumes) have been rectified, and Flight Lieutenant Cunningham's parachute would have opened. You will also appreciate this applies to any Martin-Baker ejection seat using this release mechanism'.*

Mr Butter replied on 10 April:

> *'I have now shared your e-mail with the Health and Safety Executive legal team and also Martin Baker's legal representatives for the sake of completeness. I do not at this stage require copies of the videos from you'.*

By not requesting the evidence, Mr Butter avoided the obligation to share it with the company. At no point did he ever ask for it. Mr Mark Brookes of Clyde & Co. wrote to me on 16 May 2018:

> *'The Company have been made aware of the twelve RAF training films which you have now obtained and would like to invite you to attend the business premises at Chalgrove. The Company have the ability to transfer both 16mm and 35mm film onto DVD and they would like you to use this equipment to convert your footage on to DVD, while you are present on site to observe the process. Please could you confirm if you are able to attend the business premises at Chalgrove for these purposes? If so we can then seek to make arrangements to identify a convenient date for the visit'.* [139]

I replied saying I had already paid the supplier to convert the films and

---

139 E-Mail Brookes/Hill, 16 May 2018 16:37.

they were on a single DVD. Mr Brookes replied on 23 May 2018:

> 'I have liaised with the Company and they have asked whether it is possible for you to make a copy of the DVD, which the business will pay for and if needs be can collect from you? If this is not possible then the Company would invite you to attend their premises, at Denham, so that they we can make a copy of the DVD there'.[140]

On 12 June, Clyde's Roger Cartwright asked me to send the DVD to Mr Bill Deeves, the Group General Counsel and Company Secretary.[141] I did so that day, included a draft of this book, and copied Clyde & Co. Mr Deeves acknowledged receipt on 11 July, saying the company had, as I requested, read the draft and had *'no technical comments'*.[142]

*

Returning to my exchanges with Mr Butter, on 20 April I wrote again:

> *'Instructions from Martin-Baker were not to torque load the Drogue Shackle nut. These were reiterated in Special Information Leaflets (SIL) 704 and 704A (2011/2013), and reflected in a Civil Aviation Authority Emergency Mandatory Permit Directive.*
>
> *Following release of further information yesterday by MoD, I note maintenance procedures called up when embodying Routine Technical Instruction/Hawk/059D state "tighten locknut to 50 lbf-in". This remains current. It would seem MoD is under no obligation to pay heed to a SIL, or any other bulletin, and ignores them even after a fatality. I am sure the Judge would have viewed matters differently had she known this'.*

Receiving no reply, I advised the Attorney General, who told the HSE to respond. On 17 May, in an e-mail clearly blind copied to the Attorney General, I was criticised for not providing the films - which Mr Butter had said he did not require. (While ignoring that MoD had not disclosed the same evidence). He added:

> *'The content of the films and the fact that they date back to 1959 and relate to earlier Mk4 ejection seats, and not the Mk10B seat, has satisfied me that the films do not undermine the safety of the conviction'.*

This indicated he had not watched them, or not understood the content. As he had not asked for them, and because other similar ones exist, I

---

140 E-mail Brookes/Hill, 23 May 2018 14:11.
141 E-mail Cartwright/Hill, 12 June 2018 17:12.
142 E-mail Deeves/Hill, 11 July 2018 09:48.

cannot say if he was referring to the same films. I replied:

'Had the servicing instructions shown in the film been followed, the shackles would have disengaged. This is nothing whatsoever to do with Martin-Baker. MoD is required by the Secretary of State to retain and maintain the information the films impart - especially that which is common to the Mk10B seat. (An aspect your e-mail seems to ignore). MoD is also required to disclose such evidence to you. Have you told the Judge it did not?'.

He did not reply.

*

On 4 June 2018 Mr Butter wrote again, copying Lincolnshire Police and Mrs Justice Carr. He repeated his claim that I had not provided evidence that MoD knew of the risk caused by over-tightening. He was replying to an e-mail that had included another copy of the evidence as an attachment. In my reply I finished:

'Why have witnesses not been interviewed? The current situation is you have rejected their evidence as irrelevant, without hearing it. This misled the judiciary'.

Both of us had copied the correspondence to the Judge, but on 14 June her clerk advised:

'None of the e-mails are being forwarded to or considered by the Judge or any other legal professional'.

This is called plausible deniability.

*

For the sake of completeness, I submitted a Freedom of Information request to the HSE on 1 December 2018:

'On 28 September 2016, I submitted evidence in the form of a technical paper to Helen Duggan, HSE Communications Manager, asking that it be forwarded to your investigators. Ms Duggan acknowledged receipt the same day. I updated it on 7 November 2016.

- *When was this forwarded to your investigators?*
- *When was it shared with or disclosed to the Defence, Crown Prosecution Service, Lincolnshire Police and judiciary?*
- *What action was taken to verify this evidence?*

The basis of the charge against Martin-Baker Aircraft Company was that they did not provide information to the Ministry of Defence.

- *Who in MoD was meant to be the recipient at the time in question (1990/92)? The positions held is sufficient.*
- *Was he/she interviewed, or the records of his/her decisions examined?*
- *At what point did MoD's claim that it could not find this information become 'was not provided', and how was this determined?*

The HSE did not reply. On 13 and 19 December I sought acknowledgement of receipt. This was eventually provided on 24 December, saying my request would be dealt with *'in due course'*. The Freedom of Information Act requires a response within 20 working days of receipt. (Not *acknowledging* receipt). That is, by 31 December. Allowing some leeway for the time of year, I sought progress on 14 January 2019. Receiving no reply, on 22 January I asked the Information Commissioner's Office to intervene. On 26 January, the HSE was given until 11 February to comply. Its reply was unsatisfactory:

'This was sent on 11 October 2016'.

Which does not address the more comprehensive update.

'This was shared with the Defence on 15 December 2017. HSE is not obliged to share with judiciary'.

This omits that Mrs Justice Carr had instructed the HSE to provide her, *'if appropriate'*, with copies of any material provided to the defence.[143] In other words, the HSE did not think it appropriate to disclose to the Judge evidence proving its allegations false.

The last four questions were answered with:

'We do not hold information that answers this question'.

That is, there is no record of the evidence being reviewed or witnesses interviewed. The Service Inquiry only said (wrongly) that *'evidence suggests'* MoD was unaware of the potential failure mechanism associated with tightening the nut. The HSE went further, alleging Martin-Baker never provided the information. (MoD did not demur). It now admits to having no evidence supporting its allegation. The police are required to gather sufficient evidence that will persuade the Crown Prosecution Service a prosecution is likely to succeed. And retain it. What standards apply to the HSE?

*

---

143 E-mail from Rebecca Collins, clerk to Mrs Justice Carr, 26 January 12:59.

In February 2017, the East Midlands Branch of the Institution of Occupational Safety and Health had hosted a lecture. The speakers were Mr Butter and a partner in the law firm Howes Percival, the latter setting out what the law defines as 'aggravating factors', including:

- Cost cutting at the expense of safety
- Deliberate concealment of the illegal nature of an activity
- Poor health and safety record
- Falsification of documentation

All applied to MoD, not Martin-Baker. They were recurring and had been notified to the proper authorities by the RAF Director of Flight Safety throughout the 1990s, and reiterated by the Nimrod Review in 2009. No action has been taken.

Lincolnshire Police

I noted earlier that police investigators were seriously misled. Now, I want to record the actions of their senior officers.

Suspecting my evidence had been buried by the HSE, on 20 May 2017 I sent it to the newly appointed Chief Constable, Bill Skelly, asking that it be accepted into evidence. He was obliged by law to pursue evidence of a crime, and to forward my evidence and his assessment to the HSE. I received no reply.

On 22 May 2018 I wrote to the Police and Crime Commissioner for Lincolnshire, Marc Jones, emphasising:

'Martin-Baker Aircraft Company were sentenced in February 2018. Almost immediately, exculpatory evidence emerged in the form of an MoD video disproving claims made by MoD, which had formed the basis of the Health and Safety Executive's case. Is a miscarriage of justice due to non-disclosure of this evidence a police matter?'.

Mr Jones replied on 25 May, saying he had asked Mr Skelly to reply. He added I should direct my question of a miscarriage to Mrs Justice Carr. I had already done so, without reply. (On 8 March 2019 I asked her again. She replied saying *'the sentencing process is complete'* and she had *'no further role'*. A standard tactic, also used by MoD. Repeatedly ignore correspondence, then eventually reply that it is too late).

I asked Mr Jones to whom, then, should I take evidence of the Judge being misled, as such an offence does not die with sentencing. That is, if it comes to light someone else is guilty, that party can still be pursued even in the face of another's guilty plea. He did not reply.

On 6 June 2018 Mr Skelly confirmed my evidence had been submitted to the HSE - but did not say when. He *'noted'* I had *'received a full response from the Health and Safety Executive'*. One assumes he was told this lie by the Executive.

*

On 13 June 2018 I formally reported two violations:

- *'Undermining evidence was not disclosed by MoD'*, and;
- *'When it was uncovered, it was not investigated'*.

In his reply of 27 June, Mr Skelly repeated the HSE's claim that my evidence had no bearing on the case. Assuming he read what he was replying to, this meant he was content with his officers, Coroner and Judge being misled. He urged me to make my representations to the HSE. That is, it could judge its own case.

I repeated my formal complaint on 30 June 2018:

'That this exculpatory evidence was not disclosed by MoD to yourselves, the Health and Safety Executive or courts; and that this perverted the course of justice. Also, that without interviewing witnesses the Health and Safety Executive misled the courts by claiming evidence it had not taken was irrelevant - again perverting the course of justice'.

I asked Mr Skelly if he was prepared to copy me with the advice he had received from the HSE. I offered myself for interview. He did not reply.

I tried once more, and on 3 December 2018 Lincolnshire Police told me I was *'not entitled'* to report such crimes, because I was:

'Not subject to the conduct alleged nor are you acting on behalf of any person who is entitled to make a complaint. Any further concerns should be directed towards the Coroner'.[144]

I lodged a formal complaint about this decision with the Independent Office for Police Conduct. They ruled against me, repeating the above reasoning - I was not affected, so could not report a crime. I asked for this to be reviewed, and received the same reply, twice. This flew in the face of, for example, the Crown Prosecution Service's advice that one can expect the police to take all reports seriously; at the very least taking

---

[144] Letter IX/00698/18 from Chief Inspector P. Baker, Deputy Head of Professional Standards, Lincolnshire Police, 13 December 2018.

a statement and providing a Crime Reference Number. Most police forces confirm this on their websites.

The police had now changed their minds, twice. First, I was told to approach the Judge, then the HSE, and now the Coroner. If they had been consistent, pointing me to one of these authorities, citing procedure, then I would accept that. But to fob me off in this manner was a quite deliberate act of avoiding duty. At each stage the police confirmed they knew of my correspondence with the HSE, and what the HSE said in reply. When conducted properly, this is liaison. Otherwise, it is conspiracy to obstruct justice and corruption.

*

Having refused to consider my evidence or investigate the offences I reported, on 11 February 2019 I unexpectedly received a letter from Assistant Chief Constable Shaun West. Plainly, he had been told of my recent Freedom of Information questions to the HSE, outlined above. He claimed the HSE had *'primacy of this investigation'*, distancing his force from the matter. Yet the Convening Authority, in the Service Inquiry report, stated the police had primacy.[145] Did ACC West's concern arise out of nervousness over the HSE's admission that it holds no information supporting its allegations, despite advising the Judge and police it conducted a full review?

On 26 February 2019 I submitted this book to Lincolnshire Police, asking that it be taken into evidence. And there matters stand. Justice for Sean awaits legal authorities doing their duty. Just as he was killed by MoD failing to do *its* duty.

*

The following is self-explanatory, and I discuss the recent case I mention in the Addendum. On 26 January 2022 I wrote to the new Lincolnshire Chief Constable, Chris Haward:

'In 2017-19 I exchanged correspondence with your predecessor and various senior officers. Briefly, I submitted evidence of repeated systemic failings by MoD that led directly to the death of Sean Cunningham. This included video evidence that proved the Health and Safety Executive's allegation against Martin-Baker Aircraft Ltd was false. I asked for this to be taken into evidence and reviewed.

Please be advised the same written evidence has now been the subject of a recent

---

145 Service Inquiry report, paragraph 1.6.9.

decision by the North West Wales Coroner...

At the Cunningham Inquest in 2014, MoD assured the Lincolnshire Coroner, Stuart Fisher, that breaches of its regulations identified in the Service Inquiry report had been/were being dealt with.

On 20 March 2018, Corporal Jonathan Bayliss, a Red Arrows engineer stationed at RAF Scampton, was killed at RAF Valley, Anglesey in Hawk XX204. The Service Inquiry report listed at least twelve repeat breaches. Plainly, they had not been addressed. At the Inquest in November 2021, North West Wales Coroner Ms Katie Sutherland confirmed MoD had admitted these breaches, thus directly linking the two deaths. Significantly, Sean's death was itself a recurrence (and I sent the evidence of this to ACC Shaun West on 26 February 2019).

It is a requirement that investigating authorities and prosecutors review new or fresh evidence, even after the case has been dealt with in court. These recent developments again call into question the prosecution of Martin-Baker, providing further proof that (a) evidence submitted to the Health and Safety Executive and Lincolnshire Police was exculpatory, and (b) MoD's admitted breaches were the root cause of both Sean and Jonathan's deaths.

That Martin-Baker, for commercial reasons, pleaded guilty, should not detract from these facts. But if one were of a mind to raise this issue, it would be wise to consider the sentencing remarks of Mrs Justice Carr, who set out eight distinct reasons why the company was not culpable.

I therefore request that Lincolnshire Police re-open enquiries into the death of Flight Lieutenant Cunningham. In particular, that it examines MoD's failure to correct its admitted breaches, and reconsiders the written and video evidence in the light of MoD's admission to Ms Sutherland.

Also, that the actions of the Health and Safety Executive be examined; primarily the misleading of the Courts, by omission and commission, and their prosecutorial misconduct.

My hope is that, while MoD's refusal to meet its obligations meant that, despite many warnings, the opportunity to save both Sean and Jonathan was missed, further offences and deaths may be prevented. I would be grateful if you would acknowledge receipt, provide me with a Reference, and advise what action is being taken'.

Mr Haward replied on 3 February, saying it was not *'necessary or justified to take primacy for the investigation from the HSE'*. That is, it could continue to judge its own case.

24: A pattern of behaviour

Mike Harland

Mr Harland was a British Aerospace employee, flying as a Weapons System Officer in an RAF-owned Tornado GR4 (ZA554). MoD contracted British Aerospace to undertake all maintenance and upgrade work beyond squadron capability at RAF Marham, with RAF technicians, supervisors and managers working alongside the company.

On 14 November 2007, while conducting a post-maintenance flight test at 400 knots and 5,900 feet, involving an inverted flight check, the pilot, Mark Williams, heard a loud bang followed by cockpit depressurisation. He recovered the aircraft to straight and level flight, but realised the rear canopy was shattered, with both Mr Harland and his seat missing. It transpired the seat had not been ejected, but fallen out.

The Board of Inquiry was convened by MoD's Test and Evaluation Support Division, formerly MoD's Directorate of Flying. The Air Accidents Investigation Branch was a full member.

The Top Latch Plunger, holding the seat in position until fired, had been incorrectly engaged. The assembly consists of a plunger element and a spring-loaded spigot running through its centre. (Figure 13).

At the Inquest, on 19 October 2010, Mr Williams confirmed aircrew were expected to check this locking mechanism during pre-flight checks. But, he understood it was only necessary to ensure the spigot was correctly aligned. That, he and up to 40 other civilian and military Tornado aircrew he had canvassed did not know to also check the end of the plunger was flush with its housing.

> *'I had not been aware that there was a second part of the check. Training had not revealed that 32 years of misunderstanding'.*

Only two people he asked knew this. But two is enough to indicate Martin-Baker provided the correct instructions. (In the same way groundcrew from Lossiemouth and Coningsby knew how to fit the Drogue Shackle nut correctly).[146]

The Board stated at paragraph 49a:

> *'The lack of clear instruction in the maintenance procedures may have led to*

---

[146] https://www.theguardian.com/uk/2010/oct/20/raf-tornado-seat-locking-checks-inquest

the raised inner piston going unnoticed and hence uncorrected'.

This claim was repeated at the Inquest. In fact, the instructions state the following checks must be carried out <u>three</u> times during maintenance and installation of the seat:

- *Ensure the indicating dowel pin is flush with plunger face.*
- *Ensure the plunger face is flush with edge of plunger housing.*

This is what Martin-Baker provided to customers:

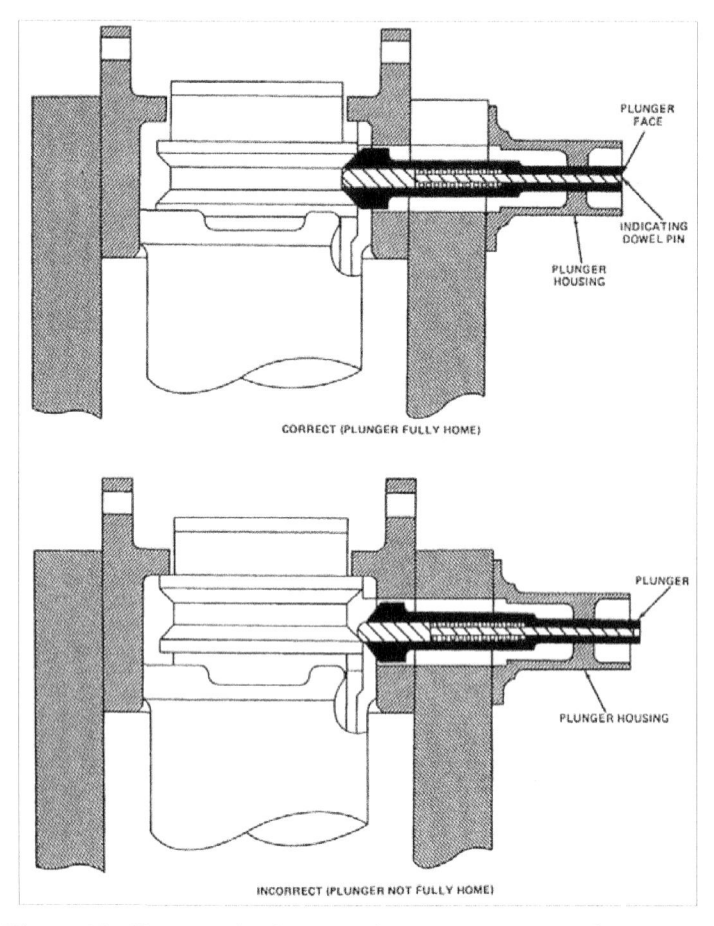

Figure 13 - Top Latch Plunger, showing correct and incorrect assembly. *(Martin-Baker)*

The earliest evidence I have of this information being provided is 1963, 44 years before the accident.[147] Once again, MoD is proven to have lost or ceased disseminating vital safety information.

At paragraph 44b, the Board stated:

'The Board identified a previously unknown condition where the inner piston could be raised to a position that prevented the correct alignment of the Top Latch Plunger, but allowed all other seat connections to be made. A raised inner piston must have existed at the time of the accident. The Board concluded that the cause of the accident was that the Top Latch Plunger was incorrectly engaged in the Top Latch Window, as a result of a raised inner piston, which led to the rear ejection seat not being locked to ZA554'.

It will be clear the Board did not *'identify'* this condition, and it was not *'unknown'*. What it identified, but omitted, was the procedures used by the RAF did not ensure serviceability or functional safety; but that Martin-Baker's did. Again, precisely what happened on XX177.

*

The reader will recall that modification 02198B introduced the Gas Shackle into Tornado seats, eliminating the risk of shackle pinching. Many of the Board's recommendations implied problems with 02198B. However, in his remarks Chief of Materiel (Air) Air Marshal Sir Kevin Leeson said:

'I do not consider that modification 02198B is deficient in meeting accepted design requirements...and therefore do not accept the recommendation for further review of modification 02198B. A correctly applied Top Latch Plunger check should have identified an incorrect fitment state.

He continued, confirming this was:

'Evidence of shortcomings with our aircrew and groundcrew training, which appear to have existed for many years, leading to a partial understanding...'.

He characterised the seat design as *'highly successful'*. That, it was fit for purpose *'so long as its fitment is taught and practiced correctly'*. He confirmed that design changes to make the assembly fool-proof might add *'complexity which would compromise its functioning when required'*.

At the XX177 hearings the Health and Safety Executive and MoD took the opposite view. Neither mentioned these remarks - perhaps because they demolished the case against Martin-Baker.

---

147 Martin-Baker Review #10 (1963) 'Important - Personal Security'.

In 1963 Martin-Baker had reported *'exhaustive tests'* had been carried out proving the Top Latch mechanism entirely satisfactory. But, crucially, *'operation must be fully understood in order to lock the seat to the aircraft correctly'*. Thus, it can be seen that for many years the company warned customers that safety of the seat was a two-way street, and suitable training was required.

Reinforcing this, significant other recommendations were accepted, such as:

- The training and authorisation of armament personnel should be reviewed.
- The 6-monthly armament recertification exam should be reviewed to ensure engineering standards and practices are maintained. A practical element should be considered.
- Industry and MoD personnel should be made fully aware of their responsibilities with respect to, and understanding the working practices of, both organisations.
- The airworthiness audit trail is maintained.
- Ejection seat training is unambiguous.

*

The degree of commonality between the two accidents is truly startling, the most obvious being:

- MoD denied having information that it possessed.
- MoD procedures included a 'test' that gave false assurance.
- Loss of corporate knowledge and poor training.
- Failure to heed Martin-Baker.

And why did the XX177 Service Inquiry have access to an engineer's report associated with the Top Latch, but the Harland Inquiry did not?

The Tornado ZA554 investigation was completed in January 2009, almost three years before Sean was killed by the same failings. Moreover, its detailed examination of modification 02198B is further evidence that MoD was aware the Gas Shackle eliminated the risk of pinching. (Because Air Marshal Leeson would not say the modification was satisfactory without confirming what it was meant to do and the effect it had on any extant risks).

Flight Lieutenant Simon Burgess

Sean would probably have survived had Disturbed Systems Testing been carried out. I looked for trends.

Simon Burgess was well-known. He had previously ejected from Tornado GR1 ZA403 on 24 January 1991 over Iraq and been held prisoner until the end of the war. When freed, he quietly returned to his duties. On 13 February 1996, he was taking off in Hawk XX164 at RAF Valley, when the aircraft rolled uncontrollably. He ejected at 150 feet, but the 110° angle of bank was outwith the design parameters for successful ejection. There was insufficient time for the main parachute to develop. The aircraft continued its roll and crashed.

Again, the similarities with XX177 are unmistakable. During a Non Destructive Testing procedure groundcrew had to disconnect another system, in this case aileron linkages, to gain access. When finished, Disturbed Systems Testing was not carried out, and the aircraft was not returned to its operational state. The ailerons were left disconnected from the flying controls. The procedure was not documented properly. The Board of Inquiry's recommendations amounted to - *implement regulations*.

*

On 27 April 2011, the Hawk fleets were grounded due to another aileron control incident caused by maintenance error. If I may take you back to Chapter 11. In October 2011, 22 Group assessed the probability of maintenance error at no more than once a year. The following month Sean was victim of such an error. The probability was then assessed as unchanged.

I shall stop there. If I hadn't mentioned this chapter was about Tornado ZA554 and Hawk XX164, you might think most of it about Hawk XX177.

25: Prevention is easier than the cure

Orchestration

Sean's death could be said to be the first major accident since the new Military Aviation Authority (MAA) got its feet under the table. Yes, there had been others between April 2010 and November 2011, but it could argue that it had not had enough time to bring influence to bear. (I'm being kind. Such organisational 'resets' are the bureaucratic way of hiding past failures. *It wasn't us guv. It was the last lot.* But it's the same people only with different post titles, and meant to be implementing the same regulations).

Nevertheless, in the 19 months after its formation the MAA conducted safety reviews of each aircraft. The scale of the task meant it drafted in 'experts' to carry out the reviews. However, the obvious weakness is that they were not independent, and some were manifestly not expert. They were a product of the failed system condemned so vehemently in the Nimrod Review, a different end of the same snake. On MoD's own admission, all had failed in their legal and moral duty to report serious wrongdoing. In the face of gross and systemic failures, they gave largely glowing reports. For example, the first question to any project team or operator would (or should) be *show me your Safety Case is valid*, because the invalidity of the Nimrod Safety Case was the Nimrod Review's headline failure. But at least Nimrod had one - albeit invalid.

This failure of duty should have been the focus of the Service Inquiry and subsequent legal action. But it was skimmed over in one brief passage, and not mentioned at all in subsequent legal proceedings. And there was nothing about why a series of very senior officers had signed to confirm they had personally verified there *was* a valid Safety Case. Nor has there been any acknowledgement that XX177 was a recurrence.

*

It is also important to understand what else was happening in the same period. Three major campaigns were running in parallel, conducted by the same members of the public:

- Chinook ZD576 (Mull of Kintyre, 1994, 29 killed)
- Hercules XV179 (Iraq, 2005, 10 killed)
- Nimrod XV230 (Afghanistan, 2006, 14 killed)

The main common denominators were (a) the aircraft were not airworthy, (b) false declarations had been made that they were, and (c) MoD had serially lied to Inquiries and families.

Precisely the same evidence was being presented in each case, to the same people - Ministers, legal authorities and the media. The links were clear, but all abetted MoD's primary aims - to compartmentalise, and protect the guilty by maliciously blaming others.

But in 2008 Oxford Coroner Andrew Walker, and his Wiltshire colleague David Masters, rejected MoD's position in the Nimrod and Hercules cases. This gave hope, and in 2009 Mr Haddon-Cave upheld my evidence, previously rejected at Ministerial level, that the failings were systemic. When the Mull of Kintyre Review was announced the following year, Lord Alexander Philip accepted the same evidence that the Chinook was not airworthy.

Each time, MoD capitulated. But it refused to change. More have died, and each time the same evidence has explained why. And each time MoD has choreographed the political, legal and media response, shutting down any discussion about common factors and causes.

*

To mark the tenth anniversary of the publication of the Nimrod Review, in October 2019 the MAA hosted an event dubbed 'Strengthening Air Safety - The Defence Aviation Environment Conference 2019'.[148] Its aim was to 'assess advancements in air safety and the future challenges that lie ahead'. The keynote speaker was Sir Charles Haddon-Cave. The MAA later said:

'Sir Charles was emphatic that the principles he identified in the report have stood the test of time and still hold true today. He also emphasised that the lessons of the Nimrod Review have relevance across all domains and not just aviation'.

And therein lay two problems: Sir Charles did not *'identify'* the failings, he reiterated them; and MoD continues to ignore the failings. Let's get back to basics. Whose damning evidence prompted the Nimrod Review, and hence the formation of the MAA? Would it not have been better to invite the author of this evidence to speak?

*

---

148 https://www.gov.uk/government/news/strengthening-air-safety-the-defence-aviation-environment-conference-2019

Earlier, I described the extent to which the Coroner and other legal authorities were misled. This rendered it almost impossible to explain to them how the accident could have been avoided. Undoubtedly, implementing Defence Standard 05-125/5 would have removed the primary root causes, ensuring the seat remained serviceable and that both it and the aircraft had a valid Safety Case. This was not an error of omission. It was a conscious policy. Sean was killed by this flat refusal to implement airworthiness regulations.

Despite the threat of disciplinary action, those aware of these failings had a moral obligation to speak out. Those with formal airworthiness delegation had a legal obligation. So, too, the police and Health and Safety Executive. When notified, they turned a blind eye. To whom, then, can aircrew and public turn to for assurance? The Ministry of Justice will not say, content with conduct that falls far below expected standards. These are big issues, in the hands of inadequates.

By withholding the Service Inquiry report until after the Inquest, MoD thinks it has cleverly avoided censure. Its senior staff were told the likely outcome of their policies (death), the deaths occurred, and they continued with the policies. They silenced those who had reported the failings and predicted the consequences. To illustrate attitudes before and after the accident, in 2003 MoD confirmed to the Permanent Under-Secretary that its policy was to discipline staff who refused to make false record regarding airworthiness and financial probity.[149] It reinforced this in 2013 when advising Ministers to uphold disciplinary action.[150] The making of false record is a key component of this case.

In such a climate, how often can a company, or junior MoD staff, be reasonably expected to step up? Industry knows MoD will do what it wants, regardless of regulations or the law. It can only try to adapt. But to what extent? Here, attempting to articulate how to assemble the Drogue Shackle *in situ* is an impossible task, because one cannot explain how to complete the job by ensuring serviceability. It would be remiss to issue partial instructions. Damned if you do, damned if you don't.

Unchecked savings, heaped upon waste

In 1991, MoD's Equipment Accounting Centre criticised as *'untenable'* the

---

[149] Letter (Reference redacted) from MoD to PS/US of S, 23 April 2003.
[150] Letter 04-06-13-145804-002, 1 July 2013, from Mark Bailey, Defence Equipment & Support Secretariat.

RAF's *savings at the expense of safety* policy.[151] Likewise, the RAF Director of Flight Safety in 1992, citing that year's 25% cut to the support budget; the second of three such consecutive cuts.[152] Without mentioning these, in 2009 Mr Haddon-Cave criticised 20% savings imposed at a political level in the period April 2000 - March 2005 (4% per year), implying this was the worst he knew of.

Contrast this with the claim by a former Survival & Airborne Delivery Integrated Project Team Leader:

> 'Under my leadership we reduced the operating costs of ejection seat support by over 40% through a transformation programme, competing a prime contract with an in-house bid. This programme resulted in three Chief Executive commendations'.[153]

This *'transformation'* took place between April 2005 and December 2006. Was the impact of the gradual 4% per year cuts assessed before making such a potentially damaging 40% cut in the following 21 months? Surely any investigator would consider this relevant, when the basic problem, and almost every engineering recommendation by the Service Inquiry, related to not implementing mandated support policy?

The 'savings' were achieved through centralised seat servicing, and extending the explosives life and seat maintenance period. One effect was that there was no seat bay at RAF Scampton in which to service the seat. Moreover, it seems no-one appreciated the need for a valid seat and aircraft Safety Cases, with safety further marginalised when the team's Safety Manager was diverted to business management. Importantly, the new policy placed greater distance between MoD and Martin-Baker.

*

In his Executive Summary to the 1998 Nimrod Airworthiness Review Team (NART) report, the RAF Director of Flight Safety, Air Commodore E. J. Black, said:

> 'The findings endorse those of various flight safety surveys over the past two years and highlight low manning levels, declining experience, falling morale and perceived overstretch generally as the driving concerns that impact directly on the Nimrod Force's ability to meet its operational task safely. The

---

151 Letter Acs EAC1E/AirTC, 19 February 1991.
152 Chinook Airworthiness Review Team report, August 1992.
153 https://www.linkedin.com/in/lxx-fxxxxxx-15916928

Force is attempting to sustain historical levels of activity with far fewer personnel and a smaller proportion of serviceable aircraft, with all the attendant hazards to safe aircraft operations. The majority of airworthiness concerns and observations tended to be linked to one central theme, i.e. the conflict between ever-reducing resources and stable or increasing demands; whether they be operational, financial, legislative or merely those symptomatic of keeping an old aircraft flying. The pressures that ensue from reducing resources call for highly attentive management, closely attuned to the incipient threat to safe standards, if airworthiness is to be safeguarded'.

These concerns and warnings were dismissed by the Air Staff as:

'...uninformed, crewroom level, emotive comment lacking substantive evidence and focus'.[154]

A briefing to the Assistant Director of Information in October 1998, from the same quarter, expressed:

'...regret that some of the content (of NART) does tend to reflect crewroom gossip/whinges rather than factual data'.

In 2011 Angus Robertson MP asked the Secretary of State to provide a copy of this briefing. Philip Hammond MP replied that after an *'extensive search'* it could not be found.[155] Yet, it had been provided to Mr Haddon-Cave in 2009, who reiterated the *'cuts, change, dilution and distraction'* arising from:

'A shift in culture and priorities towards business and financial targets, at the expense of functional values such as safety and airworthiness'.

Three years later the XX177 Service Inquiry noted that the officer who approved the Routine Technical Instruction was more concerned about how long it would take, rather than verifying the process.

Prosecutorial misconduct

This is typified by:

- Mis-stating or withholding key facts
- Ignoring factual evidence
- Suppressing evidence favourable to the accused

---

154 Nimrod Review, page 360.
155 Letter D/SofS/PH PQ05549Y/is, 20 January 2012.

- Failing to disclose exculpatory evidence
- Misleading the courts
- Assuming prejudicial facts not in evidence

All apply to this case. Specifically, to the Health and Safety Executive, aided and abetted by MoD, Lincolnshire Police, and elements of the media. All are encouraged by vague and ambiguous rules, extensive discretionary authority with no transparency or accountability, and being permitted to judge their own case.

The Health and Safety Executive had a duty to seek to understand the evidence, but did not even try. Infinitely worse, it told the Judge it had reviewed it, only to later admit it held no records of any such review. It ignored root causes and there being no proper authority for the seat to be in XX177. It initiated a selective prosecution while overlooking serious offending by MoD. Its evidence was contaminated, compromised, and corrupt. It built a myth that, if the 'missing' bulletin had landed on an MoD or RAF desk in 1990-92, the accident would have been avoided. It has, in effect, said Martin-Baker should have stepped in and carried out the tasks MoD refused to do, free of charge. It regards these matters as *'non-delegable'* by Martin-Baker, ignoring (a) that they more than met their contractual, legal and moral obligations, and (b) they cannot be expected to underwrite a build standard that does not correspond with the Master Record Index.

It further claimed that MoD's admitted training failures did not undermine the case against Martin-Baker. MoD abetted, by claiming it had no corporate knowledge of what the training had entailed - despite the answer, in the form of MoD publications and films, being readily available. The cessation of this training, and refusal to follow it, created a huge hole in each slice of the Reason Model. If these failures had not occurred, Sean Cunningham would be alive. In what way does that not undermine the Prosecution case?

This behaviour was notified to the Judge, Crown Prosecution Service, and Ministry of Justice. None have taken action, despite knowing that the HSE refused to critically examine exculpatory evidence, and failed to comply with discovery and other obligations. These legal authorities <u>knowingly</u> administered injustice, allowing the guilty to remain free.

*

The Service Inquiry set out in excruciating detail serious violations by Service and civilian personnel, but did not so much as hint at failure by

Martin-Baker. The Military Aviation Authority (MAA) then provided a Panel member as the main prosecution witness. If it had thought to, the Defence only needed to ask one question. *Do you agree with the report you contributed to, or the Prosecution's version of events?*

Unspoken disagreements exist. MoD admitted serious violations, Martin-Baker were prosecuted. MoD is content with the Mk10 seat design, the Health and Safety Executive is not. These differences were set aside to further a joint pursuit of Martin-Baker. The Executive saw an easy target and potential kudos. MoD saw a way of diverting attention from its own offences, using the Executive to run interference.

A reasonable person might think something wrong with a system that permits the Prosecution to knowingly repeat untruths *ad infinitum*. Should the Judge not step in and ask *Are you ever going to make an accusation supported by evidence?* In fact, there is no obligation on the judiciary to speak out. Defendants have an unfettered right to plead guilty, knowing that they are not. That may be acceptable if the only harm is to their coffers and reputation, but here there is a greater and continuing harm. Failure to eliminate root causes makes recurrence more likely, placing aircrew and passengers at greater risk. So too the public - especially relevant given the nature of the Red Arrows' activity.

This behaviour was unethical, improper and unjust, serving only to protect MoD. What did the prosecution achieve? Nothing. Matters have got worse, because the truth has been concealed and aircrew falsely assured. Failures have been compartmentalised. The official position is that the problem was in Hawks, and has been dealt with. The killing of Corporal Jon Bayliss in 2018, with more than a dozen common factors, proves nothing could be further from the truth.

Possible explanations

What was MoD like at the time of the alleged offence? In 1990 we didn't have personal computers or even answering machines. In Air Systems, we had one secure fax machine shared between numerous Directorates. It was in a locked room three floors below, where the sun didn't shine and nobody knew your name. Occasionally, someone would wander round and ask if you were (insert post title). In my evidence to the Nimrod and Mull of Kintyre Reviews, I cited an example whereby the policy letter at the root of airworthiness failings took 14 months to reach us in August 1988 - from a nearby RAF office in Shaftesbury Avenue.[156] I can easily envisage a Martin-Baker drawing office employee faxing a heads-up, only for it to disappear into this black hole. Knowing MoD already had the information, no reply = no problem.

The period also saw a moratorium on recruitment. Swiftly, our Directorate was carrying over 60% vacancies in specialist airworthiness posts. When lifted, none of the new recruits had worked on aircraft or equipment, or would recognise the significance of a technical bulletin. One can imagine how much work was <u>not</u> being done. Actually, you don't need to. A raft of fatal accident reports provides adequate evidence.

In June 1991 the HQ Modifications Committees were shut down. We heard a whisper and a party of us, including the Chairman, descended on our Radio Modifications Committee registry in St Giles Court, London to retrieve files before they disappeared. There was no sign of life, or filing cabinets.[157] To move staff is one thing. To remove a large truck full of cabinets without telling the data controller is quite another. It shows intent. No notification had been issued to industry, or even MoD staff, telling them of any new department or address - because there wasn't one. An entire airworthiness data repository, and control and oversight function, disappeared overnight. Did the same happen on ejection seats?

In 2002, six years after opening, MoD's new procurement HQ in Bristol suffered a multiple server failure. In my case, a single document from a 6-month old backup was recovered. The only saving grace was our instinctive distrust of assurances this could never happen, meaning older hands had kept hard copies. Why the distrust? You tend to be suspicious when the IT contractor tells you their contract excludes the concept of us dealing with anyone outwith MoD. *You want to e-mail a company in Basildon?* They laughed and walked away.

My point is, most MoD staff today would have great difficulty proving receipt of a technical bulletin last week, never mind 20 years ago. But that doesn't mean it never existed. The information may have arrived in a different form. It may have been sent to an empty office, or put in the bin by the cleaner as a result of the 'clear desk' policy. (If you leave the room, clear your desk or it will be cleared for you, never to be seen

---

156 Letter D/DDSS11(RAF)/24/1, 8 June 1987.
157 'Radio' is a generic term in MoD, referring to all electronics and associated software.

again - MoD or personal property, it mattered not). In 1991, precisely such an event meant a mounting plate for an aircraft carrier guidance system, left by an RN officer beside his desk in Main Building, was binned, leaving the ship with no data link to its Sea Harriers. (The same system I mentioned earlier, that the RAF was scrapping to save on repair costs).

I'm not offering excuses, only possibilities based on experiences. If one asks the right question, of someone familiar with the system at that time, a sensible explanation will emerge. It took me a few minutes to find the information MoD claimed it never had on seat servicing. Type *'ejection seat + servicing + film'* into a search engine. Where did I get *'film'*? We sat through them for hours on end, and were examined every week for two years. Weren't allowed remotely near an aircraft until we passed every theoretical and practical exam. More prosaically, all MoD films must be copied to the British Defence Film Library in Gerrards Cross.

Call out in silence

I believe it is important to ask how the Health and Safety Executive managed to succeed, when its case was so obviously flawed. Put another way, what did it have to arrange, do, or ensure? There were three main elements:

1. A campaign of misinformation.
2. Incriminating evidence against MoD had to be suppressed, including the compartmentalisation of what were systemic failures.
3. Preventing the independent assessment of exculpatory evidence.

I have no wish, or need, to go over these matters again. But it will be clear that MoD's participation was necessary, and I have set out their role in each deceit. From perpetuating the official line in public lectures, refusing to release key papers (especially relating to airworthiness), and the Service Inquiry failing to address critical issues such as disturbed systems testing, there can be no doubt that a conspiracy has taken place to divert the course of Justice, and maliciously blame Martin-Baker.

But who else was complicit? Matters become less clear, because there is a fine line between ineptness and misconduct in public office.

The Service Inquiry was, I believe, carried out honestly. It was denied crucial information, and it is clear the report was later edited to remove embarrassing and incriminatory facts relating to MoD's systemic safety failures, that repeated those already reiterated in the recent Nimrod Review. In particular, the direct involvement of the regulatory and convening authority (the Military Aviation Authority) in the instruction to maintainers to bypass mandated regulations, is essential to understanding how MoD escaped censure.

Similarly, the initial police investigation cannot be criticised. This was a difficult and unfamiliar area to all but very specialist investigators. There is no evidence of any such person taking part, at any time. But, again, their senior officers openly flouted legal obligations, refused to review the new evidence, and now distance themselves from their officers' investigation. This misconduct demands legal attention - unlikely, if only because it would expose the ineptness of the Crown Prosecution Service.

The Coroner and Judge were misled into making errors of fact. Both made prejudicial statements against Martin-Baker, sufficient in normal circumstances to declare a mistrial. Both accepted MoD and Prosecution lies as fact, despite having been shown the truth; however in mitigation it must be said that Martin-Baker, for reasons explained, did not defend themselves. These were not verbal opinions to be weighed against the Prosecution's argument. They were actual documents and films that the Prosecution denied the existence or relevance of. The facts spoke for themselves. The Health and Safety Executive was unable to prove any of its allegations. But myself and others, such as Sean Maffett and James Jones, were able to refute every single component of the Prosecution case using irrefutable facts and hard evidence.

For its part, the Health and Safety Executive was blind to the truth, its reasoning flawed and specious. It serially misled witnesses, courts, the family, the media, and the public. It sees this consistency as a virtue, when in my opinion it should be held criminally liable. Its actions have done nothing to further health or safety of aircrew or passengers, and the only executive decisions it made relate to their grubby, repellent persecution of Martin-Baker.

*

This is a complex and obscure scandal. Few truly grasp its depth and extent, or can identify the lie. Lacking the details I have set out, one could have speculated forever about Sean's death without getting to the unimaginable truth.

The public and bereaved have a right to know this truth. But who can they trust? Not MoD, who routinely lies to families and courts. Not the legal authorities who condone this, treating MoD as a protected species and refusing to follow the evidence. And who can aircrew and maintainers trust, in the face of MoD's policy of *savings at the expense of safety*, and default position of blaming those who cannot fight back?

For their part, Martin-Baker may not have wanted to highlight MoD's faults, but MoD did that well enough on its own. While it might be acceptable to protect individuals when the main failings are at corporate level and systemic, to protect both is an abrogation of duty - and I aim this criticism squarely at Martin-Baker, the Crown Prosecution Service, Lincolnshire Police, and the Health and Safety Executive.

Demonstrably, the only dependable and honourable input to this case came from members of the public. Who, rather conveniently, are not permitted to complain about illegal acts. This has been a cover-up and a whitewash.

So, if I may, I will draw to a close by quoting a statement by Fire Brigades Union general secretary Matt Wrack, during the Grenfell Tower Inquiry in 2018. You will identify the resonances.

> *'The Inquiry is approaching issues back to front. It is self-evident the disaster was a result of the building having been altered. While there have been expert reports provided on these issues, the Inquiry is looking at the events of the night rather than the decisions which led to it. Those who made the decisions have yet to be called to give evidence'.*

I have a suggestion. Ask Jim Cunningham if he was told of the 2002 report proving MoD knew of the risk of shackle pinching in 1984, and why it was not eliminated in Hawks. Or the training films and manuals setting out the final disengagement check. Or MoD instructing its staff not to maintain the seat in accordance with Martin-Baker instructions. Do you think he would have asked some pointed questions? No wonder he was provided with minders to drag him away from the truth.

*

My investigation is over, but the matter cannot be closed. There is a simple, fundamental issue here. The regulatory authority/user (MoD) ignored Martin-Baker's instructions as to how ejection seats must be maintained. Consequently, there was a transfer of legal responsibility for issues affected by this refusal. MoD's investigation ignored this, in turn determining the direction of the police and Health and Safety Executive investigations. (Such as they were). All three were contaminated by their disregard for MoD's refusal to heed legal obligations. Culpability is crystal clear.

Safety begins at the top. The Health and Safety Executive, and its willing accomplice MoD, have failed. There is not only concrete evidence of wrongful prosecution, but an admission of serious offending by MoD. Those who have enabled and encouraged these failures must be held to account. Legal authorities must act. Had simple regulations been followed, Sean Cunningham would be alive. The solution is not more regulations. It is to do what you're already meant to be doing.

ADDENDUM - NEW DAWN FADES

26: Conduct without conscience

Following the 2018 court case, the Health and Safety Executive's Principal Investigator, David Butter, took part in the BBC programme 'Defenders UK' (Season 1, Episode 8).[158] The false allegations and generally misleading statements were repeated. Here are a few quotes from the programme, with comments:

(HSE) *'(We look at) not just the immediate thing that has happened, but what sits underneath it and why that has happened'.*

This implies the HSE looked at contributory and aggravating factors. It did not. And as I have noted, in February 2019 it confirmed it did not review or seek to verify exculpatory evidence.

(HSE) *'The important thing was to try to manage the risk'.*

The HSE implied Martin-Baker created the risk. It was MoD.

(BBC) *'That small difference (in the nut tightness) was the difference between life and death. But this wasn't a case of human error. Mechanics were simply following manufacturer's instructions'.*

The BBC and HSE knew this to be an outright lie. The technicians were told to follow contradictory instructions issued by MoD. The shackles were clamped together, with the bolt bent and new thread cut.

(BBC) *'The outcome (of the 2018 prosecution) was what the family had hoped for'.*

The programme then showed Mr Jim Cunningham's statement outside the Coroner's Court in 2014. The BBC and HSE were aware that after Crown Court hearings of 2018 he stated to the media that he was *'horrified'* Martin-Baker had been prosecuted.

(BBC) *'The RAF had their ejection seats modified so the accident need never happen again'.*

The BBC and HSE knew, but omitted, that the 'new' shackle bolt had previously been removed from the Mk9 seat with MoD's agreement, and MoD had previously rejected a 'new' Firing Handle shroud many times. They also omitted that MoD refused to remove the over-tightening risk, by rejecting the 1983 Gas Shackle design in Hawk; but fitted it in other aircraft. And continued torqueing the nut, eschewing

---

158 Broadcast 5 December 2018, repeated 20 September 2019.

the correct tools.

But the most ironic sequence shows Mr Butter demonstrating how the shackles disengage, by pulling the Drogue Shackle out of the Scissor Shackle jaws...

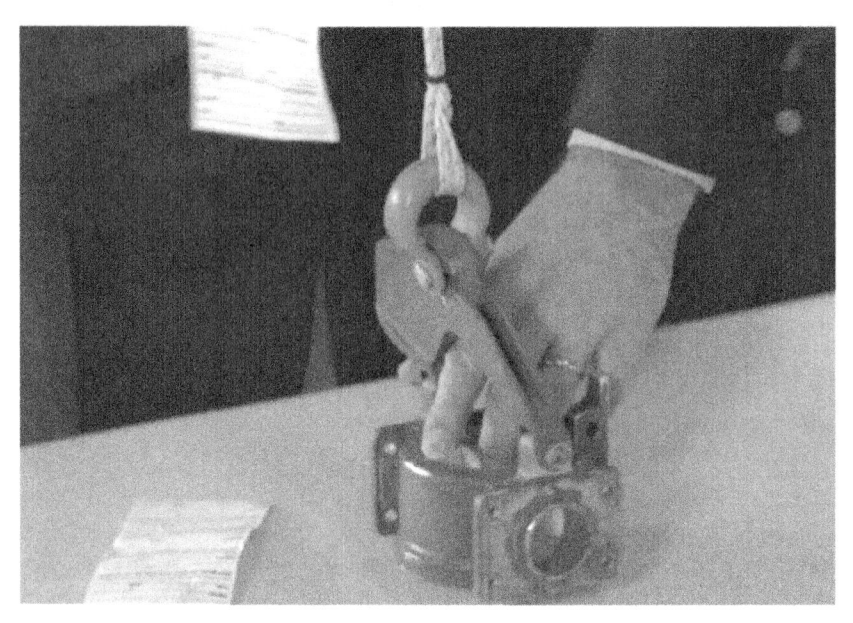

Figure 14 - HSE principal investigator demonstrating shackle disengagement *(BBC)*

His actions replicate what the servicing schedules and training films require - information he denies MoD knew about. Yet, the Civil Aviation Authority remains content that this MoD information remains available and up-to-date, as its airworthiness certification of former military jets is predicated upon it. (See next chapter).

When I notified Mr Butter of these films and publications, he claimed to have watched the former, and declared both *'irrelevant'*. So irrelevant, they taught, or confirmed to him, what was necessary to check serviceability and functional safety when performing for the BBC. Neither he nor the BBC mentioned that the RAF did not carry out this check before fitting the seat.

In effect, Mr Butter's on-camera routine is further proof that the Judge and Defence were misled. The HSE (and BBC) knew what action was necessary to prevent Sean's death. But neither revealed the whole truth. Why not? Because that would exonerate HSE's prime target (Martin-Baker), and condemn its prosecution accomplice (MoD). Both HSE and the BBC were in possession of the facts, so are party to a gross misrepresentation. Yet again, they attempted to establish an historical 'fact', without considering the overwhelming evidence that they were wrong. Who benefited?

*

On 21 September 2019, a complaint was lodged that the BBC:

- Knew that the claims made by the Health and Safety Executive (HSE) had been wholly refuted by verbal, written and video evidence.
- Misrepresented the actions necessary to ensure the ejection seat main parachute deployed correctly.

The BBC replied on 23 October 2019, refusing to address either component of the complaint.[159]

On 26 November 2019 I wrote to Ofcom, the communications services regulator. And on 16 February 2020, and again on 22 May 2020. Eventually, I received a reply in August 2020.[160] They ruled that the programme had not misled in any way. Principally, because Rule 2.2 of the Broadcasting Code requires that only news programmes must be *'duly accurate'*. In other words, any broadcaster may lie with impunity when offering programmes advertised as factual.

---

159 Letter from Deborah Lawson (BBC) CAS-5644174-XHZBV6, 23 October 2019.
160 Ofcom letter 0088053, 5 August 2020.

27: Links to the Shoreham Air Display accident

Introduction

On 22 August 2015, Hunter T7 G-BXFI crashed at the Shoreham Air Display during aerobatic manoeuvres, killing 11 bystanders and car occupants. Details were scarce until the Air Accidents Investigation Branch (AAIB) report was released in March 2017. In March 2019, at the Central Criminal Court in London, pilot Andrew Hill (no relation) was found not guilty of (a) Gross Negligence Manslaughter, and (b) Endangering the safety of an aircraft contrary to Article 137 of the Air Navigation Order 2009.

The key legal issue here is that the AAIB report was not permitted in evidence, to protect the anonymity of witnesses. One reason is that the protocols for AAIB witness statements are not in accordance with those used when taking statements that are admissible in court. Yet, the basis of the charges arose from inferences drawn from the report, so on what basis did the Crown Prosecution Service proceed?

This is not the place to rehearse the accident or the case against Mr Hill. I concentrate on what, at first, seems a peripheral matter - his ejection seat - and draw linkages to the actions of the HSE and MoD in their pursuit of Martin-Baker.

AAIB report

The report is some 452 pages long. It is in a standard AAIB format, which makes repetition unavoidable and interpretation difficult. (Much like MoD's Service Inquiry reports). It sets out selected facts, so far as they are known, without commenting on potential liability or culpability. But it unavoidably offers signposts.

The AAIB does not always establish and classify factors and causes. One must read carefully between the lines, which gives rise to many differing interpretations as to the relative importance of any given event or observation. In such a lengthy report, it is easy to lose one's way, particularly as the narrative frequently jumps between diverse subjects.

On the other hand, it does look in detail at underlying issues, such as type approval, maintenance, and the actions of authorities having oversight - in this case, the Civil Aviation Authority (CAA) and the organisers of the Air Display. Whereas, MoD tends only to look at the final act. The reason is simple. MoD is judging its own case.

It is often unclear if the authors (the various specialist Inspectors, and any Subject Matter Experts they have engaged) have been obliged to omit their expert knowledge to avoid pointing the finger, or if they have been misled. This is especially true when discussing MoD responsibilities on what used to be a Service-owned aircraft, and is relevant here.

The report makes the common mistake of confusing absence of evidence with evidence of absence. It is tainted by unverified or unexplained 'facts' being discussed in an authoritative manner, while offering no balance. Uninformed readers (primarily the police, Crown Prosecution Service and media) drew incorrect inference. My view is that this led to a politically motivated decision to prosecute, when an objective assessment of the real facts should have made these authorities think twice. Especially as even a cursory glance at the report reveals significant common factors with other fatal accidents - the death of Sean Cunningham being just one.

While Mr Hill was prosecuted, the report confirms that others erred (and probably offended) to a significant degree. No charges were brought.

Ejection seat contribution

While Mr Hill did not try to eject, the impact disrupted the fuselage under the cockpit, to the extent it initiated the ejection sequence. Only a partial ejection took place, as the damage prevented correct pressurisation of the ejection gun and firing of secondary explosive cartridges. There was therefore insufficient thrust to allow the seat to clear the aircraft in the normal manner.[161] (The primary difference between the Mk4 seat in Hunter and Mk10 in Hawk is the former is not rocket-assisted).

Nevertheless, the initial movement was enough to deploy the drogue parachutes, disengage the shackles, and release him from the seat. Both he and the seat were thrown clear. He sideways, the seat forward, its trajectory partially determined by the Drogues being snagged in trees. The fuselage then ran over it. That this aspect of the design worked as intended undoubtedly saved him.

---

161 AAIB report, paragraph 2.5.3.1.

The AAIB report studies the ejection seats in some detail, but not because of the above. It is more concerned with the cartridges being out-of-date and the bizarre servicing regime. Importantly, the report cites some of the same exculpatory evidence I presented in Sean's case. Specifically, MoD Air Publications which the HSE deny MoD had.

Suffice to say, the AAIB concluded that Hunter G-BXFI was not airworthy.[162] This was none of Mr Hill's doing. It was noted that a number of instruments were unserviceable, including the manual aileron trim gauge, which had been entered in the Technical Log as an 'Acceptable Defect' [sic]. The pilots I have spoken to deem this unacceptable, but the judge ruled the information inadmissible.

Status of 'legacy' Martin-Baker ejection seats

The Mk4HA seats in Hunters are a 1960s variant of the 1950s Mk4 seat. Production ceased in the early 1970s, and Martin-Baker have never supplied seats or components directly to civilian operators of ex-military aircraft; only to 'approved' third parties.

However, in February 2015, six months *before* the accident, the company withdrew support altogether for seats fitted to aircraft that no longer operate in their military role (including Hunters). Their reason was that the design data had become obsolete, and what was perfectly good documentation had been rendered inadequate due to lack of training in how to apply it. It can be seen that the company, unlike MoD and the Health and Safety Executive, demonstrated an appreciation of the direct link between the design pack and training. Without one, the other is useless.

A crucial factor here must be that in 1983 MoD cancelled its training contracts with Martin-Baker, and has since refused to apply the information. It is likely that, due to the loss of this income steam, Martin-Baker were unable to retain expert knowledge. By definition, this must also at some point apply to all users, past and present.[163] It is a reasonable assumption that the company's decision arose, in part, from the Cunningham case.

There is an interesting potential contradiction here. The Mk10B seats in Hawk retain the Scissor and Drogue Shackle Assembly design. Is this still

---

162 AAIB report, paragraph 2.5.4.
163 AAIB report, paragraph 1.18.15.2.

supported? Martin-Baker's stated position implies not, because the preferred and ALARP design is the Gas Shackle. If it is no longer supported, then MoD must presumably, and finally, adopt the Gas Shackle in Hawk, which it has rejected since 1984. I simply note that the old design remains in Hawk T.1.

The reader will recall the HSE's position is that the company remains responsible, regardless of the (illegal) actions of users. On 17 May 2018, it confirmed that these matters were *'non-delegable'* by Martin-Baker. In effect, the company should have stepped in and carried out the work MoD was legally obliged to do, but had abrogated.[164] It can now be seen that the HSE issued this 2018 statement in the full knowledge that Martin-Baker had withdrawn support, but failed to advise the Judge of potential implications for the Cunningham case.

Ejection seat servicing

The absence of records meant it was not possible to determine when the ejection seats had been installed in G-BXFI, only that a private individual serviced them in a 'servicing bay' he had built in his garage. This arrangement had no formal authorisation from the CAA.

The AAIB reported these basic facts, but did not address who was now responsible for maintaining the seat Build Standard, including the Safety Case. Yet, this had been a critical factor on XX177. It is unlikely that the AAIB remained ignorant of what was going on in that case, if only out of professional interest. The question then arises - if MoD investigators do not engage the AAIB, at what point must the AAIB ask to be involved if they identify common denominators? And if they do not see this as part of their role, whose is it?

*

We have already established that the Mk4 seat has the same Scissor/Drogue Shackle design as the Mk10B in Hawk. The AAIB report reinforces this with an excellent description of the mechanism, and notes:

'The current maintenance organisation...performed "bay servicing" of the ejection seats, which is required to be completed annually'.[165]

---

164 E-mail to author from HSE principal investigator, 17 May 2018 12:36.
165 AAIB report, Appendix B; and Appendix B, B1 and B2.

The Airworthiness Approval Note records at paragraphs 4.4 (Ejector Seats) and 6 (Manuals) that the seats are required (i.e. mandatory) to be serviced in accordance with AP109B-0131-12. (The Topic 5F contains Bay Servicing Schedules).[166]

I have already mentioned this Air Publication <u>five</u> times in the same context - it is what sets out the procedures for ensuring the shackles will disengage and the parachutes deploy. The content proves Martin-Baker provided sufficient information to ensure this, as it requires the maintainer to conduct a *'manual test'* to ensure shackle disengagement. Thus, my evidence is independently, if unwittingly, corroborated by the AAIB and the CAA.

A philosophical question

Some countries mandate that ejections seats in former military aircraft be inert. In the UK there is no such unequivocal policy. The CAA requires it in some aircraft, not in others. If the seat is inert, the pilot knows he will almost certainly die in a situation in which he was originally trained to use an ejection seat. But the 'plus side' is that aircraft are now cheaper to operate. Savings at the expense of safety?

Martin-Baker's position, since February 2015, is that all such legacy seats should be disabled; challenging the CAA's mandate that they should be serviceable (in Hunters). This conflict has not been resolved.

Who is the Hunter Aircraft Design Authority?

The CAA Airworthiness Approval Note for Hunter G-BXFI, dated 3 July 2008, states that it is the <u>RAF</u>.[167] (Not MoD - and as we learned earlier there is a huge difference). This would require, for example, the RAF to have arrangements in place to maintain and supply technical data for all Hunter parts. And, crucially, Safety Cases/Reports. And, because Hunter G-BXFI underwent an avionics upgrade after it was sold by MoD, the RAF would have to acquire and retain (or engage) expertise on equipment it does not use itself. An unlikely scenario, given it does not do so on equipment it *does* have.

---

166 AAIB report, Appendix L. Airworthiness Approval Note No:26172, Issue 2; to approve Hawker Hunter T Mk7 Registered G-BXFI for the issue of a Permit to Fly.
167 AAIB report, Appendix L. Airworthiness Approval Note No:26172, Issue 2, paragraph 3; issued by Civil Aviation Authority.

Moreover, the Approval Note claims all Servicing Instructions have been satisfied. You will recall that prior to April 1999 MoD's Procurement Executive issued them, and since then Defence Equipment & Support; and they are recurring. That is, it is implied there is an MoD Issuing Authority for the Instructions; or that ownership and responsibility has been formally delegated to an approved external agency along with the complete data pack - something the CAA Approval Note clearly states has not happened.

This premise, and a *'satisfactory'* Service history, is the basis of the Approval Note, issued 14 years after the RAF Out-of-Service Date. It does not address how this arrangement was, and is, to be sustained, if the RAF agreed to it, or is competent to do so. The AAIB makes no comment about the claims, yet they are central to maintaining airworthiness and to the accident itself.

In fact, this is a key issue when considering airworthiness as a whole. It is permitted, and entirely reasonable, to take increasing cognisance of historical performance. However, one must compare like with like. Any 'read across' from legacy MoD practices and achievements from even 10 years ago is entirely meaningless today, due to the policy to rundown airworthiness management, excessive outsourcing, and the very existence of the Nimrod Review. That report should have resulted in a complete reset of Safety Case justification criteria. And I would say it has, only not in so many words, and in slow time. The proof is in the series of aircraft types that have been prematurely removed from service on (lack of) airworthiness grounds. Nimrod. Sentry E-3D (AWACS). Sentinel. Gliders.

*

In an effort to verify the claims, and the current status, I submitted a Freedom of Information request to MoD on 13 September 2019. Initially, it refused to reply, but on 5 November 2019 confirmed no part of MoD was Hunter T7 Design Authority in 2008.[168] That is, the Airworthiness Approval Note for Hunter G-BXFI was invalid, based on a false premise. So, under what legitimate authority was it flying? Again, a familiar and recurring question, ignored by authorities. But why did MoD initially refuse to reply? Did my request make it aware of the CAA's claim, forcing an assessment of the implications and its possible liability?

---

168 MoD letter FOI2019/11951, 5 November 2019.

The AAIB's position remains the same, refusing to amend its report to acknowledge this anomaly.

It follows that the audit trail for transfer to the civil register is incomplete. Unless, of course, the 'privileges' mentioned include a waiver on this fundamental airworthiness requirement. These 'privileges' must be articulated and reflected in a risk assessment. A similar lack of audit trail runs through all military airworthiness.

*

The above information was submitted to the Senior Coroner for West Sussex, Penelope Schofield, on 11 October 2019. A follow-up on 6 November advised her of MoD's statement contradicting the CAA. Receipt was acknowledged, but despite a number of Pre-Inquest meetings and hearings, she has provided no feedback. It therefore remains to be seen if this evidence will be aired in court or advised to the bereaved.

The implications for the Sean Cunningham case

The HSE, via its Health and Safety Laboratory, and at the behest of the AAIB, made a significant contribution to the AAIB report (67 of 452 pages).[169] It is fair to assume it received an advance copy. On 10 September 2019 I submitted a Freedom of Information request:

'Regarding Air Accident Report 1/2017 compiled by the Air Accidents Investigation Branch into the Shoreham Air Display accident of 2015, and contributed to by the Health and Safety Executive;

a. Did the HSE reject or voice concern about any part of the report?

b. If so, may I have a copy of the comments; or alternatively an indication of the general subject?'

The HSE replied that the information was *'not held'*. I sought clarification, pointing out that the AAIB report *'...contains exculpatory evidence in another case in which your prosecution was successful. This conflict would indicate (a) comment is required on the AAIB report, and (b) the latter case is reviewed'*.

On 2 October 2019 the HSE stated it had *'provided no direct comment or*

---

169 AAIB Report, Appendix J 'Review of the 2015 Shoreham Airshow Air Display Risk Assessment' (p329-353), and Appendix K 'Review of the risk assessment sections of CAP 403' (p354-395).

concern to AAIB in response to its report'. More ambiguity, avoiding indirect comment or discussions. Once again, the HSE's sloppy approach to investigations, and failure to carry out its remit, is exposed.

Summary

The AAIB's Shoreham report was issued on 3 March 2017. On 17 May 2017 Martin-Baker pleaded not guilty in the XX177 case, unaware that the Prosecution had in its possession even more exculpatory evidence. This time, an AAIB report reproducing a CAA document citing the MoD publications containing the information Martin-Baker were accused of not providing.

Might I suggest that a procedure should be in place ensuring different investigative bodies (MoD, HSE and AAIB) are aware of issues arising from each other's work? The information is obviously recorded by each team - it is a simple matter of sharing, in the same way the police can. This would have had a profound effect on both investigations, emphasising that the accidents could not be viewed in isolation. The same basic failures occurred in both - primarily, neither aircraft was airworthy, and the seats in each were unserviceable. Immediately, that would have cast doubt on the allegations against Martin-Baker (and Mr Hill). But this should not detract from the fact the HSE knew by other means that its allegations were false.

Moreover, the reader may recall that the same failures occurred elsewhere in MoD at the same time. The Cunningham and Moray Firth Tornado Inquiries were not informed of key evidence relevant to each other's investigation. In MoD, the common denominator was the Military Aviation Authority. The Shoreham/XX177 link is less straightforward, and it is entirely possible the HSE's XX177 team was unaware. But it is now, and is required to review this fresh evidence.

28: Common factors between the Hillsborough tragedy and military accidents

This chapter may seem a digression, but I believe it important to understand it is not just MoD that is incompetent on matters of safety, and how maladministration is covered up.

*

Most readers will be aware of the Hillsborough tragedy, in which 96 football fans were killed at Hillsborough Stadium, Sheffield on 15 April 1989. The independent Hillsborough report was published in 2012, and the second Coroner's Inquests held between April 2014 and April 2016. It was ruled that the fans were unlawfully killed due to gross negligence on the part of South Yorkshire Police and ambulance services, through failing in their duty of care.

Retired Superintendent David Duckenfield, the match commander, was later prosecuted for gross negligence manslaughter. He was found not guilty in April 2019. I should say here that a senior police officer of my acquaintance, now retired, who was part of the investigation into the conduct of South Yorkshire Police, opined to me some years before that, if charged, Mr Duckenfield would be cleared, and set out the reasons. He was right on every point.

One key fact struck me. Mr Duckenfield was untrained, in the same way the Hawk maintainers were. His experience was of one end of the ground, not the whole area, so he was unfamiliar with the gates and crowd routing. This was a root cause. He was scapegoated by those who were allowed to judge their own case.

I recognised the resonance with Nimrod XV230 (2006, 14 killed). Mr Duckenfield was tasked at the last minute, as a result of another's illness. MoD's Safety Manager had been dropped into the post without training, and a QinetiQ employee told at short notice to stand in for his boss at a Safety Case meeting. Yes they made errors, but born of higher failings. Both were unfairly traduced in the Nimrod Review. Mr Haddon-Cave, supremely trained in legal matters, but not the subject in hand (MoD and military airworthiness), made many errors in his report, but luckily he was not the last line of defence against multiple deaths.

I circulated to interested parties a cross-reference index comparing Hillsborough with various military accidents. It highlighted the disparities between how the legal establishment viewed Hillsborough and military cases. Some of the issues I raised are listed below. They apply to a greater or lesser extent to Nimrod XV230, Chinook ZD576, Hercules XV179, Sea Kings XV650 & XV704, Tornado ZG710, Hawk XX177 and more.

The event, policy and background
- Savings at the expense of safety.
- Authorities failed or refused to implement mandated safety regulations.
- Lack of valid safety certification.
- The accidents were recurrences.
- Ministers and officials were given advance notice of root causes.
- The innocent were knowingly blamed.

Inquests and Inquiries
- Initial inquiries revealed the extent to which root causes were already known, and had been notified to Government and officials.
- Evidence was misrepresented or doctored.
- Relevant facts were withheld or actively concealed.
- The Inquests and Fatal Accident Inquiry (Chinook ZD576) were misled by witnesses failing to tell *the whole truth*.
- False allegations were made to divert blame.
- Legal authorities refused to pursue lines of inquiry despite being presented with evidence.
- The recommendations of Inquiries were already mandated policy.

Actions of Government and their Departments
- Ministers, media and families were misled by omission and commission, with the intention of diverting blame.
- Government and its Departments refused to accept the outcome of official Inquiries and Reviews.
- Ministers and their officials then lied to families.

- Ministers refused to take action when presented with the truth.
- Ministers and officials continued to lie when in possession of the truth.

The evidence

- Officials actively sought to prevent families uncovering evidence.
- The existence of evidence was denied, only to be uncovered by members of the public; after which authorities continued to deny its existence.
- Knowing the truth, officials/officers continued to blame the innocent.

The major differences

On Hillsborough, there were political and media calls for:

- The Inquests to be reopened.
- Honours to be withdrawn from those who committed offences.
- Legal action to be taken against the guilty.

*

Once again, the Establishment - judges, police chiefs, civil servants, Ministers and senior military officers - all fearful of openness and candour, have closed ranks to protect their own and deny the truth. When they deign to reply to legitimate questions, they repeat the same old mantra. *We have learned lessons, moved forward, and this can never happen again.* Yet time and again it does.

Their response is a classic Establishment ploy - appoint a friendly QC or judge and give him a narrow brief. On the rare occasion that one comes out against the Establishment, as Lord Alexander Philip did in his Mull of Kintyre Review, the root failings are more or less completely ignored and, in that case, the blame shifted to an anonymous junior legal officer. Only a few months later the death of Sean Cunningham proved the point. Otherwise, the vast majority of findings obligingly suit the government of the day. Hillsborough itself is further proof of this corruption of the judicial system. And like Hillsborough, it has taken independent campaigners to get at the truth.

Why does the Establishment behave in this undemocratic, dictatorial way? They are meant to be accountable, but only to themselves it seems. Victims and their families are derided, treated with contempt and indifference. Legal authorities condone, and even partake, in outright lies to protect the guilty. This is much more than them simply thinking only they are right. It is debased, depraved. It is illegal.

The conduct of the Health and Safety Executive offers a good example. Its intentions are admirable, but its implementation is high-handed, inefficient and incompetent. Witnesses, and the facts they offered, supported by verbal, written and video evidence, were ignored. The Cunningham family was treated with utter scorn, serfs with no rights. Like the emergency services at Hillsborough, it sought to cover up its failings by lying to the courts and media, without consideration for the dead or bereaved. Or, those aircrew who must follow in the wake of these failings. Even if Martin-Baker had in some way erred, this was a disproportionate and oppressive use of the law.

29: The death of Corporal Jonathan Bayliss - Hawk XX224, RAF Valley, 20 March 2018

The accident occurred as a result of a failed Practice Engine Failure After Take-Off (PEFATO). The pilot successfully ejected, just before impact. Corporal Bayliss did not initiate ejection. The Service Inquiry report was published on 10 October 2019.[170]

Command Eject

In Chapter 3, I said ejection seat operation is different if there are two occupants. The feature in question is 'Command Eject', whereby the crew have options on what happens when one of them initiates ejection. The selection is decided upon before flight, the relative experience of the passenger being a vital consideration. Command Eject is always part of the before-flight briefing, and it is crucial that pilots and passengers are fully trained and properly certified. The Service Inquiry reported that Corporal Bayliss was not.

In some aircraft, only the front seat occupant can initiate both seats. In others, both can. In many aircraft there is no Command Eject. Normally, the rear seat ejects first. This makes sense if one considers the forward speed of the aircraft. The seats immediately slow relative to the aircraft, so the front seat must be delayed to avoid hitting the rear. This sounds simple, but the concept presents a challenging design problem. For example, the rocket packs in the Mk10B seats are configured so that the front seat is propelled slightly to port, the rear to starboard.

In the Hawk, if Command Eject is OFF then each occupant ejects separately when they pull their Seat Pan Firing Handle. However, if it is ON both will eject when the rear seater pulls; the rear seat 0.35 seconds before the front. But not vice-versa. The principle here is that Hawk T.1 is a training aircraft, and the decision to eject both crew is with the Instructor in the rear.

Plainly, this logic does not stand up within the Red Arrows. The Panel concluded:

> '(Corporal Bayliss) lacked the experience to independently recognise the need

---

[170] https://assets.publishing.service.gov.uk/government/uploads/system/uploads/attachment_data/file/837526/20191007-HAWK_T_Mk1a_XX204_SI_Redacted-Final.pdf

to eject, especially as the aircraft had not suffered a technical failure and was conducting a practice procedure from which it was reasonably expected to recover. As a result, the Panel concluded that the engineer's lack of experience to independently initiate an ejection was an Aggravating Factor.

The provision of a front seat Command Eject facility may have resulted in both crew surviving, therefore the Panel concluded that the lack of such a capability was an Aggravating Factor'.[171]

The Hawk, as used by the Red Arrows, would require a different Safety Case. We know from the Sean Cunningham case there wasn't one, and now know (in 2021) there still isn't. The Service Inquiry did not go there.

*

The Convening Authority, Air Marshal Sue Gray, said in her remarks:

'Defence may have to assess if the risks associated with carrying passengers in the rear cockpit of the Hawk T Mk1 are tolerable'.[172]

But risks have to be tolerable and As Low As Reasonably Practicable (ALARP). Was this a slip, or a notion that passengers' safety need not meet the standards applied to aircrew? An uncomfortable question, but consider this...

Before the accident, Command Eject was dealt with briefly in the Red Arrows Air Safety Register, under risk RED/OTHR-I/05 'Service passenger initiating ejection of crew'. This was raised on 2 May 2012, stating:

'Worst credible outcome is pilot and passenger ejected and suffer major injuries'.

The mitigation is limited to protecting the pilot:

'Pilots always fly with Command Eject to OFF so only passengers will be ejected. Worst credible outcome becomes injuries to passenger only'.

The risk assessment assumes everyone will survive an ejection, which is nonsense. Also, that the passenger will recognise the danger and eject himself, or the pilot will issue the eject order in time. Therefore, mitigation must address the logic of the Command Eject design; and, for example, the psychology of the passenger's willingness to eject. 'Ban passengers' might be an extreme reaction, but whatever mitigation emerges it has to be covered in the Safety Case.

---

171 XX204 Service Inquiry Report, paragraphs 1.4.354 and 1.4.364.
172 XX204 Service Inquiry report, paragraph 1.6.12,

Nor does it address incapacitation; a particular problem with inexperienced passengers, who often feel ill and cope poorly with the environment. Without going into detail, there exist automatic ejection systems, but they are not designed for this scenario. But they work well, for example, when an engine fails during a vertical take-off.

It is important to note two things:

1. The Air Safety Register summary was a Service Inquiry exhibit (#67). The Engineering Risk Register is a separate exhibit. They are not linked in any way.

2. Similarly, it is MoD policy to have joint Risk Registers with industry, but the exhibit excludes any BAeS risks. (Commercial risks need not be shared). It is therefore unknown if BAeS raised a risk about the Red Arrows' concept of use contradicting the Command Eject logic.

The Service Inquiry Panel recognised the shortcomings of the register, characterising it as:

'The Delivery Duty Holder's Air Safety register appeared to be more of a personal record rather than a formal decision register'.[173]

It is difficult to think of a more damning indictment, especially post-Nimrod Review.

The Panel's recommendations

It was recommended that Air Officer Commanding 22 Group:

'Should assess the feasibility of the incorporation of a Command Eject capability into the Hawk T Mk1 that would allow aircraft commanders to initiate the ejection sequence for occupants from either cockpit seat'.

I have discussed this before. AOC 22 Group, an Air Vice Marshal, does not have the wherewithal to *'assess the feasibility'*. He does not control the funding and cannot let any contract. And even if the risk was his top priority, he has little say in the relative priority across the wider MoD, or even RAF. In fact, the most junior civil servant engineers in project teams, and even more junior ones in Service HQs, have more direct input. It is fair to ask why they had not already self-tasked, and progress recorded by the Service Inquiry.

The Service Inquiry made eight recommendations relating to AOC 22

---

173 XX204 Service Inquiry, page 1.4.105.j

Group. Five of these were outwith his gift. He would have to track down someone outwith his command willing (and able) to do the work. If, as is likely, this was unfunded and unendorsed, these staff would be unable to proceed until it was. (Part of the above self-tasking).

In her remarks, Air Marshal Gray *'agreed'* with the recommendations. But that does not constitute formal endorsement, provide funding, or grant approval to proceed. To the person tasked with constructing the case for expenditure it is meaningless, except as a minor footnote in the submission. This highlights, again, that important operational capability and safety matters are routinely facilitated and managed at a very low level. The system depends on these juniors being willing to apply common sense and engineering judgment - and their ability to fight back when told to desist. Experience tells me such people are few and far between.

Common factors with XX177

It is always illuminating to compare accident reports for recurrences. In fact, no investigation, Coroner's Inquest, or Fatal Accident Inquiry can be said to be valid without it. I note the following factors, which were notified in evidence to the North West Wales Coroner on 17 October 2019, one week after the Service Inquiry report was published:

- No common training objectives.
- Lack of continuation training
- Failure to record training.
- Improper recording of maintenance actions.
- Red Arrows adopted different procedures to other squadrons.
- Safety Case not appropriate to function and use.
- Risks not tolerable and ALARP.
- Inaccuracy and ambiguity of publications.
- Quality Assurance and authorisation failures compromising Air Safety.
- Perceived pressure arising from overworked aircrew/increased tempo/lack of resources (especially Air Safety staff).
- The accident was avoidable.

And... failure by MoD to disseminate publications and information.

The Coroner did not reply.

I shall stop there. There is too much to discuss in the Bayliss case, and I do not wish to detract from Sean Cunningham's death. It is enough, for this book, that I have drawn the direct links. The cause of death in both cases was MoD's systemic failure to implement mandated regulations.

Rather, I have published a sequel:

'A Noble Anger - The Manslaughter of Corporal Jonathan Bayliss'.

ISBN: 979-8-8342-7923-5

Glossary of terms and abbreviations

ALARP	As Low As Reasonably Practicable
BTRU	Barometric Time Release Unit
CSDE	Central Servicing Development Establishment, RAF Swanton Morley.
CSE	Central Services Establishment, MoD Llangennech.
CPS	Crown Prosecution Service
DA	Design Authority. Holds the master drawing set, whereas a Design Custodian holds secondary masters.
HSE	Health and Safety Executive. Part of the Department of Work and Pensions.
lbf-in	Pound-Force Inch - a measurement of torque.
MAA	Military Aviation Authority. Formed in April 2010 after the Nimrod Review, grouping together some airworthiness functions; but not resurrecting those shut down in the early 1990s.
MBAL	Martin-Baker Aircraft Limited
MoD	Ministry of Defence
MoD(PE)	MoD (Procurement Executive). Split into the Defence Logistics Organisation and Defence Procurement Agency in 1999, while cancelling some functions. Later recombined as Defence Equipment & Support, losing more functions.
MP	Member of Parliament
NART	Nimrod Airworthiness Review Team, and its report of September 1998.
NAS	Naval Air Squadron
RAF	Royal Air Force
RAFAT	Royal Air Force Aerobatic Team
RTI	Routine Technical Instruction
RTS	Release to Service. The Master Airworthiness Reference, stating the limitations within which Service regulated flying may be conducted.
SIL	Special Information Leaflet
TART	Tornado Airworthiness Review Team, and its report of November 1995.
UTI	Urgent Technical Instruction

Printed in Great Britain
by Amazon